Image Processing and Data Analysis with ERDAS IMAGINE®

Image Processing and Data Analysis with ERDAS IMAGINE®

Stacy A. C. Nelson, PhD
Siamak Khorram, PhD

CRC Press is an imprint of the
Taylor & Francis Group, an **informa** business

CRC Press
Taylor & Francis Group
6000 Broken Sound Parkway NW, Suite 300
Boca Raton, FL 33487-2742

© 2019 by Taylor & Francis Group, LLC
CRC Press is an imprint of Taylor & Francis Group, an Informa business

No claim to original U.S. Government works

Printed in Canada on acid-free paper

International Standard Book Number-13: 978-1-1380-3498-3 (Hardback)

This book contains information obtained from authentic and highly regarded sources. Reasonable efforts have been made to publish reliable data and information, but the author and publisher cannot assume responsibility for the validity of all materials or the consequences of their use. The authors and publishers have attempted to trace the copyright holders of all material reproduced in this publication and apologize to copyright holders if permission to publish in this form has not been obtained. If any copyright material has not been acknowledged please write and let us know so we may rectify in any future reprint.

Except as permitted under U.S. Copyright Law, no part of this book may be reprinted, reproduced, transmitted, or utilized in any form by any electronic, mechanical, or other means, now known or hereafter invented, including photocopying, microfilming, and recording, or in any information storage or retrieval system, without written permission from the publishers.

For permission to photocopy or use material electronically from this work, please access www.copyright.com (http://www.copyright.com/) or contact the Copyright Clearance Center, Inc. (CCC), 222 Rosewood Drive, Danvers, MA 01923, 978-750-8400. CCC is a not-for-profit organization that provides licenses and registration for a variety of users. For organizations that have been granted a photocopy license by the CCC, a separate system of payment has been arranged.

Trademark Notice: Product or corporate names may be trademarks or registered trademarks, and are used only for identification and explanation without intent to infringe.

Library of Congress Cataloging-in-Publication Data

Names: Nelson, Stacy A. C., author. | Khorram, Siamak, author.
Title: Image processing and data analysis with ERDAS IMAGINE® / Stacy A.C. Nelson and Siamak Khorram.
Description: Boca Raton, FL : Taylor & Francis, 2018.
Identifiers: LCCN 2018010060 | ISBN 9781138034983 (hardback : alk. paper)
Subjects: LCSH: Remote-sensing images--Data processing. | Image processing--Digital techniques. | Imagine (Computer file).
Classification: LCC G70.4 .N45 2018 | DDC 621.36/78--dc23
LC record available at https://lccn.loc.gov/2018010060

Visit the Taylor & Francis Web site at
http://www.taylorandfrancis.com

and the CRC Press Web site at
http://www.crcpress.com

Contents

Acknowledgments xi
Authors xiii
Introduction and Overview xv

**1. Acquiring Data: EarthExplorer, GloVis, LandsatLook Viewer,
and NRCS Geospatial Data Gateway** 1
Overview 1
Acquiring Remotely Sensed Data 4
    Learning Objectives 5
I. Finding and Downloading Data in EarthExplorer 6
II. Automated Method of Importing EarthExplorer and Creating a
Multi-Band, Layer Stack, Image in ERDAS IMAGINE 27
III. Finding and Downloading Data in GloVis 34
IV. Displaying Raster Data and Creating a Multi-Band Image in
Esri ArcMap ArcGIS for Desktop 39
V. Displaying Raster Data and Creating a Multi-Band Image in
Quantum Geographic Information Systems 45
Review Questions 47

2. Introduction to Image Data Processing 49
Overview 49
Introduction to Digital Image Processing Application 52
    Learning Objectives 52
I. Obtaining Required Data in EarthExplorer 53
II. ERDAS IMAGINE Graphic User Interface 56
    Exploring ERDAS Help Documents 57
    Setting Up Workspace Preferences 59
    Opening Images 60
    Getting Data Information 62
    Band Combinations 62
    Create a False Color Composite Display Band Combination 64
    Linking Images in Multiple 2D Viewer Windows 66
    Opening Multiple Images 66
Review Questions 68

3. Georectification 69
Overview 69
Image Preprocessing—Georectification 70
    Learning Objectives 70

v

Rectifying Image of Schenk Forest Using Polynomial Regression
and Rubber Sheeting 71
  I. Polynomial Regression 71
    Display Images—Start Two Viewers (File | New | 2D View) 71
    Record Ground Control Points 74
    Compute Transformation Matrix 80
    Resample the Image 81
    Verify Rectification 82
  II. Rubber Sheeting 86
    Display Images 86
    Start Geometric Correction Tools 87
    Ground Control Points 87
    Compute Transformation Matrix 89
    Resample the Image 90
    Verify Rectification 91
  Review Questions 93

4. Orthorectification 95
  Overview 95
  Image Preprocessing—Orthorectification 95
    Learning Objectives 96
    Getting Started—Orthorectification 96
    Defining Camera Properties (Interior Orientation) 99
    Selecting Ground Control Points 109
  Creating an Orthophoto 114
  Viewing the Orthophoto 115
  Review Questions 116

5. Positional Accuracy Assessment 117
  Overview 117
  Positional Accuracy Application 119
    Learning Objectives 119
  Calculating Error in the X and Y Directions 119
  Root Mean Square Error 120
  Total Root Mean Square Error 120
  Euclidean Distance 121
  Review Questions 121

6. Radiometric Image Enhancement 123
  Overview 123
  Radiometric Enhancement Application 124
    Learning Objectives 124
  Performing Radiometric Enhancements 125
    Loading Stretched and Non-Stretched Images 125

Contents vii

Understanding the Stretch...127
ERDAS IMAGINE and Lookup Tables (LUT Values) 129
Adjusting the Stretch..133
Making Finer Adjustments..137
Review Questions 146

7. Spatial Image Enhancement...147
Overview..147
Spatial Image Enhancement Application 148
Learning Objectives...148
Spatial Image Enhancements 148
Initiate a 3×3 Filter 149
Initiate a 5×5 High Pass Filter 150
Initiate a 5×5 Low Pass Filter...150
Open the Non-Direction Edge Filter Tool Dialog Window 150
Open the Image Degrade Tool Dialog Window 151
Review Questions 151

8. Image Digitizing and Interpretation...153
Overview..153
Image Digitization and Interpretation Application.........................153
Learning Objectives...154
Polygon Creation...155
Polyline Creation...161
Manual Drawing...164
Changing the Display Properties of the Digitizing Results.............165
Review Questions 166

9. Unsupervised Classification 167
Overview..167
I. Unsupervised Classification Application 169
Learning Objectives...169
Obtaining the Required Data (Review) 169
Creating an Image Subset of the Cloud-Free Areas........................172
Initiating the Unsupervised Classification 176
Unsupervised Classification Application Approach 177
Creating a Reference Image Subset 182
Assigning Class Categories to the Unsupervised Classification......184
Compare 188
II. Unsupervised Classification in Esri ArcMap ArcGIS
for Desktop..190
III. Unsupervised Classification in QGIS 199
Review Questions 206

viii

Contents

10. Supervised Classification 207
Overview.. 207
Supervised Classification Application.. 209
   Learning Objectives... 209
Obtaining the Required Data (Review) 210
Initiating the Supervised Classification 212
   Supervised Classification Application Approach—Creating a
   Signature File... 213
   Tips for Creating the Supervised Classification 218
   Compare Using Swipe Tool 219
   Compare Using Image Difference Operation (Change Detection) 221
   Note on Image-To-Image Change Detection Comparisons.............. 228
Review Questions 229

11. Object Based Image Analysis 231
Overview.. 231
Object Based Image Analysis Classification Application 232
   Learning Objectives... 232
I. Feature Project Setup Procedure.. 233
II. Feature Extraction Procedure.. 237
Examine the Object-Oriented Classification in ERDAS IMAGINE...... 245
Review Questions 246

12. Additional Image Analysis Techniques 249
Overview.. 249
Additional Analysis Techniques Application.. 250
   Learning Objectives... 250
I. Create a Land-Only Image.. 251
   Initiating the Classification Procedure 252
      Create a Thematic Image Recode.. 253
      Create a Binary Image Mask to Remove Water 256
II. Create a Normalized Difference Vegetation Index 258
   Create a Normalized Difference Vegetation Index 258
      Classify the Normalized Difference Vegetation Index.................. 260
III. Create an Impervious Surface Map.. 262
   Create an Impervious Surface Map (Remove Vegetation) 262
   Classify the Impervious Surface Map... 264
   Recombine Classification Components 266
   Combining Output Layers in Model Maker 266
Review Questions 271

Contents

ix

13. Assessing Thematic Classification Accuracy 273
    Overview 273
    I. Assessing Thematic Classification Accuracy Application 274
        Learning Objectives 274
    II. Recoding the Supervised Classification 275
    III. The Accuracy Assessment Procedure 277
    IV. Accuracy Assessment Report Generated from
    ERDAS IMAGINE 284
    Review Questions 287

14. Basics of Digital Stereoscopy 289
    Overview 289
    Basics of Digital Stereoscopy Application 289
        Learning Objectives 289
    Configuring the Stereo Analyst module 290
    Making a Digital Stereo Pair 293
    Creating Anaglyph Images for Export 298
    Delineating in Stereo 300
    Review Questions 306

Appendix: Answer to Chapter Review Questions 307

References 321

Index 323

Acknowledgments

The authors wish to thank several individuals who have contributed materials, developed application examples, and explored many workarounds, all in an effort to find the most meaningful and useful ways of conveying a very complex set of skills contained within this book. Georectification, orthorectification, positional accuracy assessment, and digital stereoscopy activities were adopted from original research and laboratory exercises developed by Dr. Heather Cheshire. Workaround for automated data import method for downloading and creating a multi-band composite image from the USGS EarthExplorer website in ERDAS IMAGINE was adopted from techniques explored by Ms. Melinda Martinez. Object Based Image Analysis techniques were adopted from original research and laboratory exercises developed by Mr. Kevin Bigsby. We appreciate the continual input and feedback from, Dr. Brett Hartis, Dr. David Barry Hester, Mr. William (Bill) Slocumb, Dr. Ernie Hain, Dr. Halil Cakir, Laura Belica, and students within the Center for Geospatial Analytics and the Department of Forestry and Environmental Resources at North Carolina State University.

Authors

Stacy A. C. Nelson is a professor in the Department of Forestry and Environmental Resources and a research fellow with the Center for Geospatial Analytics at North Carolina State University, Raleigh, North Carolina. Dr. Nelson received a BS from Jackson State University, Jackson, Mississippi, an MA from the College of William and Mary, Williamsburg, Virginia, and a PhD from Michigan State University, East Lansing, Michigan. His research centers on GIS technologies to address questions of land use and aquatic systems. He has worked with several federal and state agencies including; the NASA Stennis Space Center in Mississippi, the NASA-Regional Earth Science Applications Center (RESAC), USDA Forest Service, as well as various state level agencies. He is active in several professional societies.

Siamak Khorram is a professor of remote sensing and image processing. He holds joint faculty appointments at both University of California at Berkeley and North Carolina State University in Raleigh. He is also the founding director of the Center for Geospatial Analytics at North Carolina State University and a member of the board of trustees at the International Space University (ISU) in Strasbourg, France. Dr. Khorram is a former vice president for academic programs and the first dean of ISU, as well as a former chair of the ISU's Academic Council. He was an ASEE fellow at Stanford University and NASA Ames Research Center, California. Dr. Khorram has extensive research and teaching experience in remote sensing, image processing, and geospatial technologies, and has authored 226 publications including two textbooks. He has served as the guiding/major professor for over 30 master's and doctorate graduate students. He is a member of several professional and scientific societies. His graduate degrees were awarded by the University of California at Davis and University of California at Berkeley.

Introduction and Overview

Introduction

Remote Sensing is the science of acquiring information about an object without physically coming in contact with it. Remotely sensed data, acquired in a raster format, is collected by a diverse array of passive and active sensors mounted on aircraft (including airplanes, helicopters, unmanned aerial vehicle (UAV)/drones, and balloons) or spacecraft (usually satellites, but also Space Shuttle missions and the International Space Station). Data used in image processing operations generally include three main types of data; imagery (or image data), ground-based or reference data, and other ancillary data types on an as-needed basis. The image data may consists of photogrammetric (air photos) or satellite imagery, captured in various digital formats. The aerial photography may consist of black and white photos, color photos, or color infrared (CIR) imagery. The black and white and color photos are typically not multispectral data and consist of a single static image. While this data may be useful for high-resolution observations over large or small areas, it typically is not well-suited for digital image analysis. However, it does provide useful ancillary data for reference analysis and accuracy assessment of land cover classifications. Usually, the CIR imagery will consist of multiple images/bands, captured simultaneously, that represent surface reflectance information in the visible (blue, green, and red) and near infrared ranges of the electromagnetic spectrum (EMS). This range of the EMS is very useful for capturing the strong reflection of vegetation within this region. The electromagnetic spectrum is illustrated in Figure 0.1.

Satellite or aircraft imagery may be acquired either passively or actively. Passive imagery data is formed from a sensor that typically captures varied amounts of the sun's radiation that is reflected or emitted from objects on the earth's surface within the visible, near infrared and shortwave infrared or thermal infrared portions of the EMS. As such, these passive sensors may be referred to as optical or infrared sensors. Four basic types of sensor imagery result from passive sensors; visible, infrared, multispectral, and hyperspectral imagery.

Active imagery data is formed from a sensor that uses an antenna to actively generate pulses of energy that are returned to the sensor. The returned pulses are then captured to form an image. Active sensors are typical of RAdio Detection And Ranging (RADAR) and Light Detection And Ranging (LiDAR) sensors. LiDAR data can be collected from air, Airborne Laser Scanner (ALS) or on the ground, Terrestrial Laser Scanners (TLS).

The second main types of data used in image processing operations are ground/reference data. Ground data, as the name implies, is typically

xv

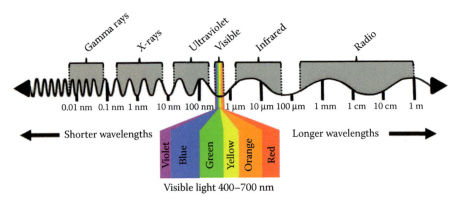

FIGURE 0.1
Electromagnetic spectrum.

collected on the ground at the same time of sensor overflight (data acquisition) or within a time period, which can reasonably be assumed that no major changes on the ground have occurred between the collection of the ground data and the sensor overflight. This type of data may include many forms, such as Geographic Positioning System (GPS) data, topographic maps, vector and GRID datasets, and even high-resolution aerial photographs (such as, black and white, color, and CIR). This data is typically used for sensor calibration, geometric correction, land cover validation, sensor design, and so on.

The final main types of data used in image processing operations are auxiliary data. Auxiliary data types may consist of various types of data, including ground data and additional files that are associated with the image data. These additional files may include the image header file information, a metadata file, pyramid or graphics files, or even topographic maps, GRID and vector data that match the image area. The metadata and header file information are of critical importance, as these files contain specific data description information, information on the platform, sensor, data format, acquisition date, bit depth, projection information, and so on. When downloading data from internet web-portals, always check to make sure a metadata file is included with the download (typically available as a .TXT or .XML file). Should this file be missing or not included with the data, the only way to be certain about any questions you have about the data would be to contact the originator of the data.

Remotely sensed image form or raster data are useful for digital image processing and are typically arranged in a two-dimensional matrix of cells or picture elements that are commonly referred to as *pixels*. This two-dimensional matrix of cells is organized into a specific number of rows and columns that will be dictated by the type of sensor the image was acquired from. Each pixel contains the reflectance data of specific ranges within the EMS that the sensor was designed to capture. Each pixel also corresponds

Introduction and Overview xvii

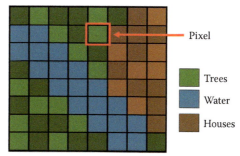

(a) Real landscape (idealized)   (b) Digital raster image of real landscape

FIGURE 0.2
(a) Represents an idealized real landscape comprised of trees, water, and houses and (b) represents a raster image captured of the idealized real landscape consisting of a matrix of cells (or pixels) organized into rows and columns.

to the particular location on the Earth's surface from where the image was collected. Additionally, the ground spatial resolution of the sensor corresponds to the individual pixel size that each image is comprised of. For example, the spatial resolution of the multispectral scanner (Operational Land Imager, or OLI) aboard the Landsat 8 satellite is 30 m. This means that of the thousands of pixels that make up a 185 km (across track) by 170 km (along track) image or scene, each pixel will have a spatial resolution corresponding to a 30 m by 30 m area on the ground (Figure 0.2).

The digital raster imagery is also typically captured as a multiband raster dataset, typically referred to as multispectral data. The sensor-array records naturally occurring electromagnetic radiation (EMR) that is either reflected or emitted from areas and objects of interest on the surface of the earth within a specific range on the EMS. For example, the OLI sensor-array aboard the Landsat 8 satellite has four ranges that it records from with the visible portion of the EMS. These ranges correspond to the sensor's spectral band placements: that is in the visible range, Band 1—Ultra Blue, 0.43–0.45 µm; Band 2—Blue, 0.45–0.51 µm; Band 3—Green, 0.53–0.59 µm; Band 4—Red, 0.64–0.67 µm. The term band (short for spectral band) refers to the color ranges or *color band* on the electromagnetic spectrum (Figure 0.3).

During the latter half of the twentieth century, large-scale space programs, such as National Aeronautics and Space Administration (NASA), the European Space Agency (ESA) and the Japanese Space Agency (NASDA) were primarily responsible for the development of new remote sensing technologies. Such programs remain key sources of technological innovations. Additionally, the expansion of these remote sensing technologies has contributed greatly to the proliferation of sensors by many other national space agencies (including India, Russia, Germany, France, China, South Korea, Brazil, Netherlands, Finland, Spain, and so on) and the private sector such as DigitalGlobe, as well as others that provide high-quality data for a variety of applications.

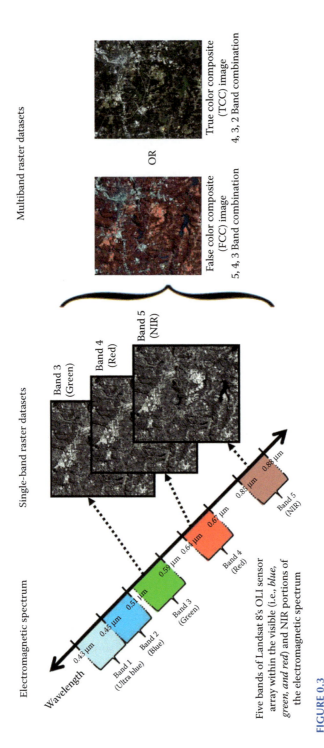

FIGURE 0.3
Multiband Raster Dataset displayed in a False Color Composite (FCC) image (5, 4, 3 Band Combination) and a True Color Composite (TCC) image (4, 3, 2 Band Combination) in relation to Landsat 8's OLI sensor-array within the visible (blue, green, and red) and near infrared (NIR) portions of the electromagnetic spectrum.

Introduction and Overview xix

Since 2008, the entire Landsat data archive has been made publically available at no cost through many US government websites (see Chapter 1 for links). The Landsat satellite program represents the longest, continuously operational land-imaging program in the world, dating back from early 1972 through the present day. The free and open access to this data provides managers, researchers, and students of remote sensing the richest, multispectral satellite data archive currently available, spanning well over four decades of repeated observation of the earth's surface on a consistent platform. Other government-based satellite data acquisition programs, such as ESA's Sentinel-2 Multispectral Imager (MSI) data products, have also become publicly available at no cost through accessible web portals.

Overview

The main goal of this text is to provide an easy to follow, step-by-step guide to processing remotely sensed digital images mostly available at no cost to users. This text is organized as a practical guide that allows the reader to *learn-as-you-go*. The principles and theories of image processing and data analysis are interjected throughout the steps of the individual operations, providing readers with a hands-on experience that demonstrates proof of concept as each application is accomplished. This exercise gives the readers the ability to quickly gain the practical experience that will allow them to easily move beyond the information presented in this text and tackle more advanced skills. Many exercises are highlighted using publically available imagery, such as NASA Landsat, Moderate Resolution Imaging Spectroradiometer (MODIS), Visible Infrared Imaging Radiometer Suite (VIIRS) data, and others. More and more sensor data types are increasingly becoming available. For example, in November 2009, ESA implemented a policy offering free, open access to Sentinel-2 MSI data to all users through the Copernicus Open Access Hub (https://scihub.copernicus.eu/dhus/#/home). This decision was followed in May 2010, as part of the ESA Earth Observation Data Policy, to provide free of charge, open access to ERS-1, ERS-2, Envisat, GOCE, SMOS, CryoSa, and all future Earth Explorer missions. ESA's Earth Online (EO) data can be accessed through the "How to access ESA Earth Online (EO) data" tutorial: https://earth.esa.int/web/guest/data-access/how-to-access-eo-data/how-to-access-earth-observation-data-distributed-by-esa.

The initial chapter of this text explores a few common US government websites for acquiring remotely sensed data. However, many state, county, and municipal agencies now offer remotely sensed data in many formats, including black and white, color, and color infrared high-resolution aircraft imagery. This text and the associated exercises focus on providing basic image processing skills, primarily utilizing Hexagon Geospatial's digital image processing package, known as ERDAS IMAGINE (http://www.hexagongeospatial.com/products/power-portfolio/erdas-imagine). Although this text also highlights the use of a few other image processing packages, ERDAS IMAGINE has long been considered the industry standard in remotely sensed image processing and data analysis software. The collection of exercises offered in this text, although primarily aimed at the newer user, provides skills that will assist even more advanced users, as not only a great introduction to the software, but also to the data and how these methods and algorithms are applied specifically to these types of data. This text will also provide a foundation for further advanced explorations into the world of digital image processing for practical applications, as well as highlighting certain applications in Esri's ArcMap ArcGIS for Desktop software and Quantum the GIS (QGIS) open source applications package.

Please refer to *Principles of Applied Remote Sensing* (Khorram et al. 2016) and *Remote Sensing in Springer Briefs in Space Development* (Khorram et al. 2012) for more detailed information on remotely sensed data, platforms, sensors, principles, and applications.

Disclaimer: This text was compiled through the authors' ability to access the latest available technologies, resources, and software packages. We are aware that this technology is ever evolving and hope to be able to provide updates through a companion website and/or updated editions, as any changes become available that may affect the workflows described in this text. Unfortunately, it is not possible to provide a demonstration for every remote sensing or image analysis software application available today. Although this text primarily uses the Hexagon Geospatial's ERDAS IMAGINE® software, as well as several duplicate demonstrations in Esri's ArcGIS Desktop, and the QGIS open sources software, the functionality of the application-based methods presented here should work similarly, regardless of the software application being used. It is the hope of the authors that the demonstrations presented in this text will be used for educational purposes and provides a robust introduction into the vast world of remotely sensed, image processing and analysis.

References

Khorram, S., F. H. Koch, C. F. Van Der Wiele, S. A. C. Nelson, and M. D. Potts. 2016. *Principles of Applied Remote Sensing*. Springer Science+Business Media. New York. p. 307. ISBN 978-3-319-22559-3. DOI 10.1007/978-3-319-22560-9.

Khorram, S., F. H. Koch, C. F. Van Der Wiele, and S. A. C. Nelson. 2012. Remote Sensing *in Springer Briefs in Space Development*. Springer-Verlag. New York. p. 141. ISBN 978-1-4614-3102-2. DOI 10.1007/978-1-4614-3103-9.

1

Acquiring Data: EarthExplorer, GloVis, LandsatLook Viewer, and NRCS Geospatial Data Gateway

Overview

Remote sensing can be characterized as being the science of acquiring any information about an object or a phenomenon on the surface of the Earth without coming into physical contact with it. Typically, with remote sensing image analysis from aircraft or satellite sensors, light energy as it relates to the electromagnetic spectrum, is captured and used to form *images*. This process is based on the spatial, spectral, radiometric, and temporal resolutions of the images and are commonly referred to as *data* in this book. The electromagnetic energy that emanates from the Earth's land surface, oceans, and atmosphere, provides a method to identify, delineate, and distinguish between these features (Barr and Barnsley 2000, Short 2010). Electromagnetic radiation (EMR) is defined as all energy that moves with the velocity of light in a harmonic wave pattern. The harmonic wave pattern refers to the waves that are equally and repetitively spaced in time. The visible light, the portion of the EMR that humans can see, is just one category of EMR. Other categories of EMR include radio waves, microwaves, RADAR, infrared, ultraviolet, X-rays, and gamma rays. For a more in-depth discussion of EMR, please see Khorram et al. 2016b.

Spectral or optical remote sensing is primarily based on detecting and recording reflected and emitted EMR. The ability to remotely sense features is possible only because every object, or material, has a particular emission and/or reflectance property (Jensen 2005). These properties can be analyzed and compiled into a spectral signature, or profile, that represents the feature and distinguishes it from other objects and materials. Remote sensors are designed to collect this *spectral* data. This data is stored in a *raster* image format. In general, a raster dataset consists of a matrix of cells, or pixels, organized into rows and columns. This row and column organization resembles a grid in structure. Each cell contains a value representing

1

information, such as temperature, brightness, or other specifications. Each raster may represent the captured electromagnetic energy from the emitted and/or reflected objects and compiled into a collective image format and resolutions tied to the sensors specifications (Lillesand et al. 2008). For example, the data of a NASA Landsat 7 (launched in April 1999) is comprised of a 30-meter spatial resolution in its spectral bands, 16-day temporal resolution, 8-bit radiometric resolution, and a spectral resolution range from 0.45–2.35 um within the EMR spectrum (Khorram et al. 2012a, Khorram et al. 2016a).

Aerial photography, or photogrammetry, usually consists of three types of image products; black and white, color, black and white infrared, and color infrared. Black and white images usually consist of lower-cost surveys, which are used to collect information over large regions (Morgan et al. 2010). Due to the lower cost of this data, and the possibility of more extensive use and coverage, multiple generations of this data type are ideal for comparing recent changes in land surface. Color imagery may represent higher-cost aerial photo surveys. However, due to the increased cost of acquiring this data, comprehensive inventories across large regions may only be limited to specific areas, such as municipalities. However, U.S. government programs, such as the National Agriculture Imagery Program (NAIP), acquire aerial imagery during the agricultural growing seasons across continental U.S. The imagery captured (typically) on three-year cycles and comprised of 3-band, natural color images (such as red, green and blue) and in some states 4-band red, green, blue, and near infrared images, since 2007. Please view United States Geological Survey (USGS) NAIP interactive coverage map of the conterminous United States for more information on availability (https://www.fsa.usda.gov/Assets/USDA-FSA-Public/usdafiles/APFO/NAIP_Covg_20150512.pdf). This data is available through the U.S. Department of Agriculture (USDA) Natural Resources Conservation Service: Geospatial Data Gateway (https://gdg.sc.egov.usda.gov/), and the USGS EarthExplorer (http://earthexplorer.usgs.gov/) websites. Infrared imagery is primarily useful for vegetation studies. This is due to the fact that vegetation is a very strong reflector of infrared radiation within the EMR spectrum (Mattison 2008).

Spectral remote sensing data may also be acquired from satellite or aircraft platforms. Satellite platforms provide the greatest capacity to cover large areas routinely and repetitively. For optical remote sensing, the imagery is acquired passively. This simply means that passive sensors record intensities of electromagnetic radiation that are either emitted or reflected from an object, typically within the visible and/or infrared portions of the EMR spectrum. However, most passively collected data are limited to energy not absorbed by the Earth's atmosphere. Passive sensors may collect data in multiple bands. Each band corresponds to a particular region, or segment, within the electromagnetic spectrum.

Acquiring and importing data into digital image analysis packages, such as ERDAS IMAGINE®, provides an opportunity to extract pertinent information for available raster data sources. ERDAS IMAGINE provides an extensive variety of tools that allow users to take advantage of spatial, spectral, and radiometric nature of the data for critical evaluation, and subsequent geospatial analyses. For that reason, the main emphasis of the practical applications in this book is focused on using ERDAS programs. Additionally, many remotely sensed data sources are now available at a reduced cost, or completely free. For example, since 2008 the entire Landsat data archive has been made publicly available at no cost through U.S. government agency portals. Landsat satellite program represents the longest, continuously operational land-imaging program in the world, since July 23, 1972, through present day.

Landsat satellite imagery as well as several other sensor and platform image data types are available for download or viewing at no charge from a number of different sources across the internet, including the following websites:

USGS EarthExplorer (http://earthexplorer.usgs.gov/)

Query and order satellite images, aerial photographs, and cartographic products through the USGS.

USGS Global Visualization Viewer - GloVis (http://glovis.usgs.gov/)

The USGS Global Visualization Viewer (GloVis) is a quick and easy online search and order tool for selected satellite and aerial data.

USGS LandsatLook Viewer (http://landsatlook.usgs.gov/)

Developed as a prototype tool to allow rapid online viewing and access to the USGS Landsat image archives at up to full resolution directly from a web browser. Landsat data is commonly available within 24 hours of reception.

NASA Satellite Data and Imagery Resources – Reverb (https://reverb. echo.nasa.gov)

Formerly known as the Warehouse Inventory Search Tool, or WIST.

USDA Natural Resources Conservation Service: Geospatial Data Gateway (https://gdg.sc.egov.usda.gov/)

This site is particularly good for finding orthorectified aerial photographs (geometrically corrected or orthophotos), NAIP imagery, or Geographic Information Systems (GIS) data. The site's well-organized search engine allows searching by county, state, bounding box, and other criteria.

Sentinel Hub's Earth Observation (EO) Browser (http://www. sentinel-hub.com/apps/eo_browser)

Sentinel Hub's EO Browser represents a newer web-based browser style resource for finding and downloading imagery. Currently, imagery is available for European Space Agency's (ESA) complete archive of Sentinel-2, Sentinel-3, Proba-V, and Landsat 5, 7 and 8 (global data coverage is provided with Landsat 8). Sentinel Hub is operated by Sinergise, a GIS IT company headquartered in Ljubljana, Slovenia.

Acquiring Remotely Sensed Data

This exercise uses the USGS EarthExplorer and the USGS Global Visualization Viewer (GloVis) websites to search for and download available satellite raster data for further analysis. The techniques employed here are routine and demonstrated through obtaining image data for the ESA's Sentinel-2 Multispectral Instrument (MSI) of the Paris, France region (File Name: L1C_T31UDQ_A010058_20170526T105518, Acquisition Date: 2017/05/26) and NASA's Landsat 8 Operational Land Imager/Thermal Infrared Sensor (OLI/TIRS) imagery of the Raleigh, North Carolina region (File Name: LC08_L1TP_015035_20160413_20170223_01_T1, Acquisition Date: 2016/04/13). However, the procedure demonstrated here will allow you to find available data anywhere in the world and for any available dates.

The Landsat 8 OLI/TIRS or simply Landsat 8 was launched on February 11, 2013, and represents the United States' continuing mission of the Landsat sensor program. The sensor's images consist of nine spectral bands (Bands 1–7, and 9) with a spatial resolution of 30 meters, one panchromatic band (Band 8) with a spatial resolution of 15 meters, and two thermal, surface temperature bands (Bands 10 and 11) with a spatial resolution of 100 meters (resampled to 30 m). Landsat 8 has a revisit cycle of every 16 days and approximate scene size of 170 km north-south by 185 km east-west (106 mi by 115 mi) (Table 1.1).

TABLE 1.1

Table of Landsat 8 Operational Land Imager (OLI) and Thermal Infrared Sensor (TIRS)

Band	Wavelength (micrometers)	Description (Representation of the Electromagnetic Spectrum)	Resolution (meters)
Band 1	0.43–0.45	Coastal aerosol (Ultra Blue)	30
Band 2	0.45–0.51	Blue	30
Band 3	0.53–0.59	Green	30
Band 4	0.64–0.67	Red	30
Band 5	0.85–0.88	Near infrared (NIR)	30
Band 6	1.57–1.65	Shortwave Infrared (SWIR) 1	30
Band 7	2.11–2.29	Shortwave Infrared (SWIR) 2	30
Band 8	0.50–0.68	Panchromatic	15
Band 9	1.36–1.38	Cirrus (IR)	30
Band 10	10.60–11.19	Thermal Infrared (TIRS) 1	100 (resampled to 30)
Band 11	11.50–12.51	Thermal Infrared (TIRS) 2	100 (resampled to 30)

Acquiring Data: EarthExplorer, GloVis, LandsatLook Viewer

The Sentinel-2 multispectral sensor is a component of the ESA's Copernicus Programme. The initial Sentinel-2A satellite was launched June 23, 2015. This was followed up by a Sentinel-2B satellite launch on March 7, 2017. The two Sentinel satellites (2A and 2B) operate in a 180° phase with the orbit allowing for global coverage of the Earth's surface every five days, or less at higher latitudes (every 10 days with one satellite). The Sentinel sensors' images consist of 13 spectral bands and operate in the visible/near infrared (VNIR) and shortwave infrared spectral range (SWIR). The spectral bands possess a mix of spatial resolutions (10 m, 20 m, and 60 m). The approximate scene size (also known as tiles) is 100 km north-south by 100 km east-west (62 mi by 62 mi). While the Sentinel-2 sensors do not possess the long Earth observation legacy available through the Landsat program, the higher spatial resolution and revisit cycles make this newly available dataset complementary to the types of global land surface investigations that Landsat has been valued for many years (Table 1.2).

TABLE 1.2

Table of Sentinel-2 Multispectral Imager

Band	Central Wavelength (micrometers)	Description (Representation of the Electromagnetic Spectrum)	Resolution (meters)
Band 1	0.433	Coastal aerosol (Ultra Blue)	60
Band 2	0.490	Blue	10
Band 3	0.560	Green	10
Band 4	0.665	Red	10
Band 5	0.705	Vegetation Red Edge	20
Band 6	0.740	Vegetation Red Edge	20
Band 7	0.783	Vegetation Red Edge	20
Band 8	0.842	Near Infrared (NIR)	10
Band 8A	0.865	Narrow Near Infrared (NIR)	20
Band 9	0.945	Water Vapor	60
Band 10	1.375	Shortwave Infrared (SWIR) - Cirrus	60
Band 11	1.610	Shortwave Infrared (SWIR)	20
Band 12	2.190	Shortwave Infrared (SWIR)	20

Learning Objectives

1. To increase familiarity with publicly available data search tools for image processing.
2. To learn to identify suitable data and acquire data based on certain project requirements.
3. To learn to import this data directly into ERDAS IMAGINE.

I. Finding and Downloading Data in EarthExplorer

With a web browser, connect to the EarthExplorer website by typing the following URL: http://earthexplorer.usgs.gov/.

If you have not registered for an account with the USGS, you may do so by clicking the "**Register**" link in the upper-right corner of the EarthExplorer browser. If you have an account already, you may log in at this time by clicking the "**Login**" link in the upper right corner of the EarthExplorer browser (Figure 1.1).

FIGURE 1.1
EarthExplorer web browser.

1. Enter the search criteria

    There are three options: Address/Place, Coordinates, and/or Date Range. The map viewer window can be clicked to define the search area for any place of interest in the world. Additionally, the range of dates of interest for obtaining available imagery may be entered. Alternatively, if you know the address, place name, or exact coordinates, you can enter those options to locate an area of interest. This exercise will search for an image of Paris, France.

    Type in the word "*Paris*" in the Address/Place field and click on the "**Show**" button. A list of available areas fitting the specified search

Acquiring Data: EarthExplorer, GloVis, LandsatLook Viewer

criteria will appear in the following list. Click on the Paris, France link under the Address/Place heading to show the location on the map and add coordinates to the Area of Interest Control (Figure 1.2).

FIGURE 1.2
Search criteria in the EarthExplorer web browser.

Notice the map viewer window places a red location balloon on the target area, and the Coordinates section of the Search Criteria updates with the coordinates of the selection. Feel free to use the **Plus** icon to the lower right of the map viewer window to zoom into the target area (Figure 1.3).

FIGURE 1.3
Coordinate section in EarthExplorer web browser.

2. Select a Date Range

Enter in the "**Search from**" in the Search Criteria for the following dates: For example, **05/01/2017** to **08/31/2017**.

Next select "**Data Sets**" below Date Range options.

3. Select "**Landsat Archive | L8 OLI/TIRS C1 Level-1**" and the "**Sentinel | Sentinel-2**" categories and click on the **Information** icon next to each available product-type to learn more about each item (Figure 1.4).

FIGURE 1.4
Product type information in EarthExplorer web browser.

Also, notice that there are other available data collections here as well. For information about these data formats, see the online help documentation or the USGS Long Term Archive website: https://lta.cr.usgs.gov/products_overview.

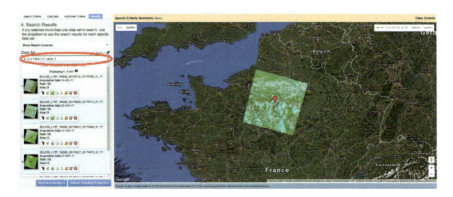

FIGURE 1.5
Available Landsat OLI/TORS data products in EarthExplorer web browser.

4. Next select **"Results"** at the bottom of Data Sets tab (Figure 1.5).

For this example, the majority of available **L8 OLI/TIRS** products (representing the Landsat 8 Operational Land Imager and Thermal Infrared Sensor) are obscured by cloud cover. This imagery would make land use/land cover image analysis very difficult. It is best to try to locate imagery within the target area with as minimal cloud cover as possible.

Next, review the available **Sentinel-2 MSI** products by changing the drop-down list from **L8 OLI/TIRS** to **Sentinel-2** (Figure 1.6).

Acquiring Data: EarthExplorer, GloVis, LandsatLook Viewer 9

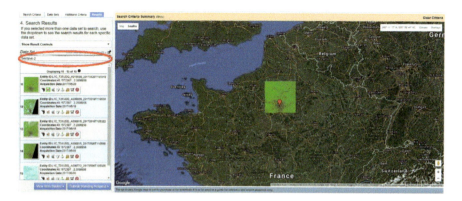

FIGURE 1.6
Available Sentinel-2 MSI data products in EarthExplorer web browser.

5. Select the image data

The displayed Search Results represent all image data within the specified date range.

Review each available image. Ideally, it is recommended to select an image (or images) that completely covers the target area (or area of interest) and has minimal cloud cover so as not to obscure the area of interest.

The **Show Footprint** icon under each available image is useful for ensuring the satellite image scene's footprint completely encompasses the area of interest (Figure 1.7).

FIGURE 1.7
Show image scene footprint in EarthExplorer web browser.

The **Show Browse Overlay** icon to determine whether any cloud cover in the scene will obscure the area of interest.

Finally, check the **Show Metadata and Browse** icon [icon] to determine the recorded image quality (option available for L8 OLI/TIRS products) and the actual amount of cloud cover in the scene to not obscure the area of interest. In Figure 1.8, Sentinel-2 image **L1C_T31UDQ_A010058_20170526T105518** meets the specified criteria.

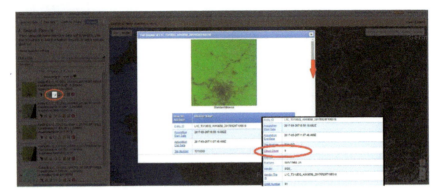

FIGURE 1.8
Show Metadata and Browse icon in EarthExplorer web browser.

6. Downloading the image data

   To download the image, select the **Download Options** icon [icon] to determine the available image products (L1C_T31UDQ_A010058_20170526T105518).

   > NOTE: You may also choose the **Add to Bulk Download icon** [icon] should you need to download more than a single image.

   The **Download Options** icon [icon] gives us several data options. For Sentinel-2 MSI data, the download options will include the following files:
   a. **L1C Tile in JPEG2000 format**
   b. **Full Resolution Browse in GeoTIFF format**

   The **L1C Tile in JPEG2000 format** file represents the MIS data you selected. The **Full Resolution Browse in GeoTIFF format** file provides a high-resolution TIFF image for further examining the scene area or for use as ancillary data.

   NOTE: For L8 OLI/TIRS products, the LandsatLook options provide high-resolution JPEG images (such as LandsatLook Natural Color

Image) for further examining the scene area or use as ancillary data. The **"Level 1 GeoTIFF Data Product"** provides the image data in the form of GeoTIFF images (i.e., LandsatLook Images with Geographic Reference). In a GeoTIFF image, the georeferencing information is embedded within the individual TIFF image files making it possible to find coordinate locations on the image.

NOTE: The USGS EarthExplorer provides a "Landsat Analysis Ready Data (ARD)" download option for acquiring Landsat 4-5 Thematic Mapper (TM), Landsat 7 Enhanced Thematic Mapper Plus (ETM+), and Landsat 8 OLI/TIRS. The ARD images are now available for the conterminous United States (or CONUS), Alaska, and Hawaii. For the CONUS, ARD images are available from 1982 to the present, and each ARD image is a 5,000 pixels-square scene. Every Landsat scene with data over a target scene contributes to the multi-date stack whether the scene is whole or partially covered for an area. While the ARD product ensures that a location may be captured across successive/historical image dates, it should also be noted that not each ARD scene will represent the same amount of spatial area coverage on the ground and any partial scene may limit the ability to develop further historical classifications in adjacent areas.

Alternatively, you may also select the **"Item Basket"** link in the upper-right corner of the map viewer browser window, or the **"Submit Standing Request"** button at the bottom of the **"Results"** tab to begin the download.

7. Unzipping the image data.

Unzipping the Sentinel-2 MSI data:

To unzip the **L1C_T31UDQ_A010058_20170526T105518.zip** file that downloads to your computer, use a freely available compression program such as WinZip, 7-zip, or any other available unzip software to unzip this file.

NOTE: WinZip, or the generic Windows unzip application, may experience an error if the file name or the directory path that the file is being unzipped from is too long. The freely available, open source, software 7-zip (http://www.7-zip.org/) typically does not experience these issues.

The unzipped file has a ***.SAFE** (Standard Archive Format for Europe) extension. Navigate to the following unzipped directory on your computer: ...\S2A_MSIL1C_20170526T105031_N0205_R051_T31UDQ_20170526T105518.SAFE\GRANULE\L1C_T31UDQ_A010058_20170526T105518\IMG_DATA. The IMG_DATA directory will contain multiple files (mostly single image files), each representing an individual sensor band and a True Color Image in JPEG2000 format (Table 1.3).

TABLE 1.3

Table of Sentinel-2 Dataset Downloaded from USGS EarthExplorer Displaying Bands Grouped According to Spatial Resolutions. The Details the Band Archive Available in the L1C_T31UDQ_A010058_20170526T105518 Dataset are Shown in This Table

File	Description (Color; Spatial Resolution)
T31UDQ_20170526T105031_B01.jp2	Sentinel-2 Band 1 (Coastal Aersol; 60 m)
T31UDQ_20170526T105031_B02.jp2	Sentinel-2 Band 2 (Blue; 10 m)
T31UDQ_20170526T105031_B03.jp2	Sentinel-2 Band 3 (Green; 10 m)
T31UDQ_20170526T105031_B04.jp2	Sentinel-2 Band 4 (Red; 10 m)
T31UDQ_20170526T105031_B05.jp2	Sentinel-2 Band 5 (Vegetation Red Edge; 20 m)
T31UDQ_20170526T105031_B06.jp2	Sentinel-2 Band 6 (Vegetation Red Edge; 20 m)
T31UDQ_20170526T105031_B07.jp2	Sentinel-2 Band 7 (Vegetation Red Edge; 20 m)
T31UDQ_20170526T105031_B08.jp2	Sentinel-2 Band 8 (Near Infrared; 10 m)
T31UDQ_20170526T105031_B8A.jp2	Sentinel-2 Band 8A (Near Infrared; 20 m)
T31UDQ_20170526T105031_B09.jp2	Sentinel-2 Band 9 (Narrow Near Infrared; 60 m)
T31UDQ_20170526T105031_B10.jp2	Sentinel-2 Band 10 (Cirrus-Shortwave Infrared; 60 m)
T31UDQ_20170526T105031_B11.jp2	Sentinel-2 Band 11 (Shortwave Infrared; 20 m)
T31UDQ_20170526T105031_B12.jp2	Sentinel-2 Band 12 (Shortwave Infrared; 20 m)
T31UDQ_20170526T105031_TCI.jp2	True Color Image in JPEG2000 format

NOTE: The individual Sentinel-2 MSI bands displayed in Table 1.3 are color-coded to match their pixel spatial resolutions. Ideally when creating a multi-band image composite that combines these single bands, care should be taken in ensuring only bands with similar spatial resolutions are used. The bands highlighted in yellow will be used for creating a Sentinel-2 multi-band image.

Further description of these Sentinel-2 MSI files can be found in the SENTINEL-2 User Handbook (https://earth.esa.int/documents/247904/685211/Sentinel-2_User_Handbook).

Unzipping the L8 OLI/TIRS data:

Use the same procedure demonstrated earlier to find and download a Landsat 8 OLI/TIRS image of the Raleigh, North Carolina region (File Name: LC08_L1TP_015035_20160413_20170223_01_T1, Acquisition Date: 2016/04/13) (Figure 1.9).

The Landsat 8 files can be unzipped in a similar fashion with a few exceptions. The "**Level 1 GeoTIFF Data Product**" will be downloaded as a zipped Tape Archive (TAR) file that has been additionally zipped a second time as a GZip file (.gz):

LC08_L1TP_015035_20160413_20170223_01_T1.tar.gz

To unzip this file, use WinZip, 7-zip, or another available unzip software to unzip this file **twice**! The first unzip of the GZip file (*.gz) will reveal a TAR file (LC08_L1TP_015035_20160413_20170223_01_T1.tar). Unzip the TAR file using the same unzip procedure a second

Acquiring Data: EarthExplorer, GloVis, LandsatLook Viewer 13

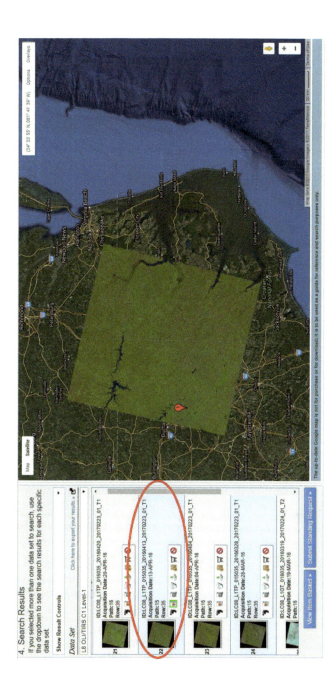

FIGURE 1.9
Landsat 8 OLI/TIRS image of the Raleigh, North Carolina region (File Name: LC08_L1TP_015035_20160413_20170223_01_T1, Acquisition Date: 2016/04/13) selected in EarthExplorer web browser.

14   *Image Processing and Data Analysis with ERDAS IMAGINE®*

TABLE 1.4

Table of Landsat 8 Dataset Downloaded from USGS EarthExplorer Displaying Bands Grouped According to Spatial Resolutions. This Table Details the Band Archive Available in the LC08_L1TP_015035_20160413_20170223_01_T1 Dataset

File	Description (Color; Spatial Resolution)
LC08_L1TP_015035_20160413_20170223_01_T1_B1.TIF	Landsat 8 Band 1 (Coastal aerosol; 30 m)
LC08_L1TP_015035_20160413_20170223_01_T1_B2.TIF	Landsat 8 Band 2 (Blue; 30 m)
LC08_L1TP_015035_20160413_20170223_01_T1_B3.TIF	Landsat 8 Band 3 (Green; 30 m)
LC08_L1TP_015035_20160413_20170223_01_T1_B4.TIF	Landsat 8 Band 4 (Red; 30 m)
LC08_L1TP_015035_20160413_20170223_01_T1_B5.TIF	Landsat 8 Band 5 (Near Infrared; 30 m)
LC08_L1TP_015035_20160413_20170223_01_T1_B6.TIF	Landsat 8 Band 6 (Shortwave Infrared 1; 30 m)
LC08_L1TP_015035_20160413_20170223_01_T1_B7.TIF	Landsat 8 Band 7 (Shortwave Infrared 2; 30 m)
LC08_L1TP_015035_20160413_20170223_01_T1_B8.TIF	Landsat 8 Band 8 (Panchromatic; 15 m)
LC08_L1TP_015035_20160413_20170223_01_T1_B9.TIF	Landsat 8 Band 9 (Cirrus; 30 m)
LC08_L1TP_015035_20160413_2017022 3_01_T1_B10.TIF	Landsat 8 Band 10 (Thermal Infrared 1; 100/30 m)
LC08_L1TP_015035_20160413_2017022 3_01_T1_B11.TIF	Landsat 8 Band 11 (Thermal Infrared 2; 100/30 m)
LC08_L1TP_015035_20160413_2017022 3_01_T1_BQA.TIF	Landsat Quality Assessment Band
LC08_L1TP_015035_20160413_2017022 3_01_T1_MTL.txt	Metadata File

time! The unzipped TAR file will reveal the multiple files (mostly single image files), each representing an individual sensor band (Table 1.4).

NOTE: The individual Landsat-8 bands displayed in Table 1.4 are color-coded to match their pixel spatial resolutions. Ideally when creating a multi-band image composite that combines these single bands, care should be taken in ensuring only bands with similar spatial resolutions are used. The bands highlighted in yellow will be used for creating a Landsat 8 multi-band image.

Further description of these Landsat 8 files can be found in the USGS Landsat 8 Data Users Handbook (http://landsat.usgs.gov/l8handbook.php).

Acquiring Data: EarthExplorer, GloVis, LandsatLook Viewer 15

8. Displaying raster data in ERDAS IMAGINE

   To display the individual Sentinel-2 MSI bands you previously unzipped in ERDAS IMAGINE, each band will need to be imported in the current data format of each file (i.e., JPEG2000 format, or file extension .jp2).

   a. Start ERDAS IMAGINE from the start menu (**Windows Start | All Programs | ERDAS IMAGINE**), or in simply type "**ERDAS**" into the Windows search box (Figure 1.10).

FIGURE 1.10
Hexagon Geospatial's ERDAS IMAGINE application package interface. (From ERDAS IMAGINE®/Hexagon Geospatial.)

   b. From the File menu tab select "**File | Open | Raster Layer**" (Figure 1.11).

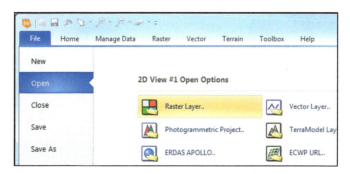

FIGURE 1.11
ERDAS IMAGINE open raster layer command. (From ERDAS IMAGINE®/Hexagon Geospatial.)

   c. Next, in the **Select Layer to Add** dialogue window that opens, navigate to the directory that you downloaded and unzipped your data to using the **Up Directory** icon and/or the "Look in" drop-down list box (Figure 1.12).

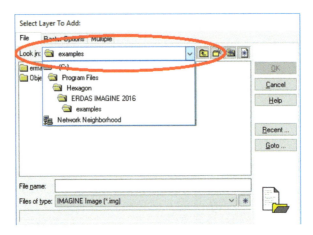

FIGURE 1.12
ERDAS IMAGINE select layer to add dialog window. (From ERDAS IMAGINE®/Hexagon Geospatial.)

d. Then in the "Files of type" drop-down list, select "**JPEG 2000**" as the file type (Figure 1.13).

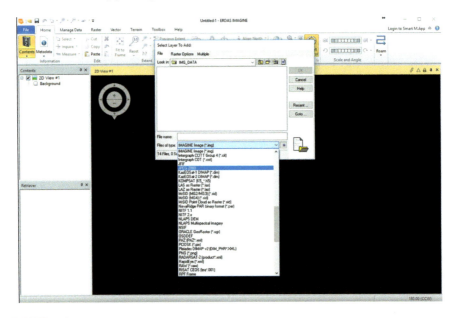

FIGURE 1.13
ERDAS IMAGINE import for JPEG 2000 (*.jp2) file format. (From ERDAS IMAGINE®/Hexagon Geospatial.)

NOTE: For the Landsat 8 OLI/TIRS files, the unzipped format will result in individual GeoTIFF files (*.tif). As a result, you would need to select "**TIFF**" as the file type to import these files into ERDAS IMAGINE.

e. Select each available JPEG 2000 file for viewing in ERDAS IMAGINE by clicking the file and then selecting "**OK**" in the **Select Layer to Add** dialogue window.

You may receive an option to create a Pyramid Layer file may appear (Figure 1.14).

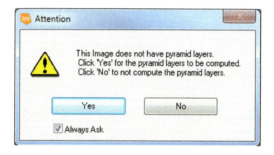

FIGURE 1.14
ERDAS IMAGINE create pyramid layers option. (From ERDAS IMAGINE®/Hexagon Geospatial.)

The pyramid file (*.rrd) is created within the same output directory as the original file(s). This reduced resolution dataset simply stores graphic information to allow the original file(s) to display quicker. Select "**Yes**" to initialize this file and "**Close**" in the Process List dialog window that follows once the operation is shown as complete.

9. Creating a Layer Stacked image (Sentinel-2 MSI data)

In order to have the flexibility of displaying band combinations from the downloaded image files (versus simply viewing a single image at a time), create a layer stack of the bands of interest or a multi-band image. The software displays properties in most image processing packages by combining up to three bands within the viewing window of a multi-band image.

As noted previously, the Sentinel-2 MSI bands come in a mixture of spatial resolutions. A multi-band image composite should only be created from the single bands that possess similar spatial resolutions. This would mean combining only the 10 m (*<u>bands 2, 3, 4, and 8</u>), only the 20 m (*<u>bands 5, 6, 7, 8A, 11, and 12</u>), or only the 60 m (*<u>bands 1, 9, and 10</u>) spatial resolution bands. (*<i>band numbers color-coded to match their pixel spatial resolutions</i>). If bands 1–12 and 8A are added to the layer stack operation in ERDAS IMAGINE, all bands in the resulting multi-band image would be resampled to the smallest pixel-meter size (such as 10 m). This would result in the loss of some spectral definition in the bands with larger, original, spatial resolutions of 20 m and 60 m, and potentially providing unreliable results if these bands are used in classification.

In creating a multiband image with the Sentinel-2 MSI bands, use the individual bands 2, 3, 4, and 8 since they all have a similar spatial resolution of 10 m.

a. First, close any images remaining in the 2D Viewer (**File | Close | Close All Layers**).

b. Next, from the file menu in ERDAS IMAGINE, click on the "**Raster**" tab. Then select "**Spectral**" from the Resolution grouping. Finally, click on "**Layer Stack**" (Figures 1.15 and 1.16).

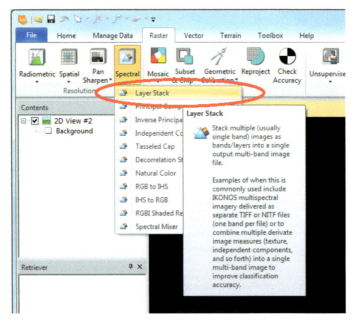

FIGURE 1.15
ERDAS IMAGINE create layer stack option. (From ERDAS IMAGINE®/Hexagon Geospatial.)

c. In the Layer Selection and Stacking dialog window, create a layer stacked image, also known as a multi-band image, composed of Sentinel-2 MSI's spectral bands (Band 2–4, and 8). Add each of the Input Files, **one by one,** as follows (be sure to click on "**Add**" to add the file selections individually to the Layer List). The first image should be added as ***Layer: 1*** since the image is comprised of multiple single layers (this should be the default option). After adding the first single layer image, all subsequent files can also be added individually as using the default (Layer: 1) option:

Acquiring Data: EarthExplorer, GloVis, LandsatLook Viewer 19

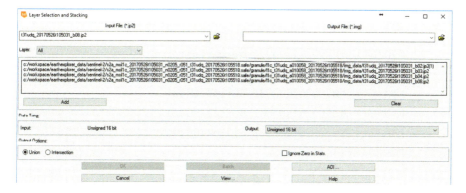

FIGURE 1.16
ERDAS IMAGINE Layer Selection and Stacking dialog window. (From ERDAS IMAGINE®/Hexagon Geospatial.)

T31UDQ_20170526T105031_B02.jp2: Layer: 1
T31UDQ_20170526T105031_B03.jp2: Layer: 1
T31UDQ_20170526T105031_B04.jp2: Layer: 1
T31UDQ_20170526T105031_B08.jp2: Layer: 1

NOTE: The new multi-band image will consist of bands 1–4. Remember that Sentinel-2's band 2 will be represented in the new multi-band image as band 1, since Sentinel-2's original band 1 was not included in the layer stack operation. This will also be the case for the remaining bands in the new multi-band image (Table 1.5).

TABLE 1.5

Comparison of Original Band Sentinel-2 MSI Bands to the Resulting Layer Stacked Multi-Band Image

Original Sentinel-2 Band	Layer Stack Multi-Band Image
Original Sentinel-2 Band 2	Multi-Band Image Band 1
Original Sentinel-2 Band 3	Multi-Band Image Band 2
Original Sentinel-2 Band 4	Multi-Band Image Band 3
Original Sentinel-2 Band 8	Multi-Band Image Band 4

It is not necessary to add the following files to the Layer Stack image for this exercise, as these are captured at different spatial resolutions:

T31UDQ_20170526T105031_B05.jp2 (20 m)
T31UDQ_20170526T105031_B06.jp2 (20 m)

T31UDQ_20170526T105031_B07.jp2 (20 m)

T31UDQ_20170526T105031_B8A.jp2 (20 m)

T31UDQ_20170526T105031_B11.jp2 (20 m)

T31UDQ_20170526T105031_B12.jp2 (20 m)

T31UDQ_20170526T105031_B01.jp2 (60 m)

T31UDQ_20170526T105031_B09.jp2 (60 m)

T31UDQ_20170526T105031_B10.jp2 (60 m)

The new multi-band image will consist of bands 1–8. Remember that Landsat 8's band 9 will be represented in the multi-band image as band 8, since Landsat 8's band 8 was not included in the layer stack operation.

Also note, the radiometric resolution of the Sentinel-2 MSI's instrument is 12-bit, enabling the image to be acquired over a range of 0 to 4095 potential light intensity values. However, images are distributed as 16-bit unsigned integers.

d. Next, name the output file: **T31UDQ_20170526T105031_B02-4_8** and change the output file type to "Files of type: **IMAGINE Image (*.img)**" from the drop-down list. Also, check "**Ignore Zero in Stats.**" under Output Options.

NOTE: The "..._B02–4_8" suffix in the file name represents a designation that the newly created file will now be comprised of the original Sentinel-2 bands 2–4, and 8 (Figure 1.17).

The output file will be saved as an *.img* file. This is ERDAS IMAGINE's native format for raster files and often the default format that the software will save files in after operations that require creating a new file.

e. Select "**OK**" in the Output File and Layer Selection and Stacking dialog windows.

Acquiring Data: EarthExplorer, GloVis, LandsatLook Viewer 21

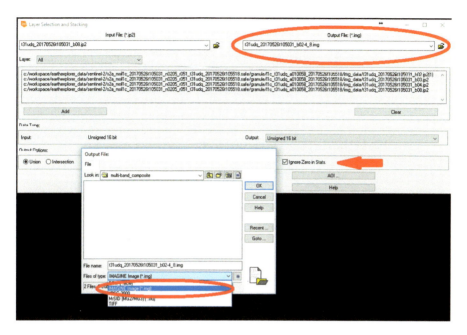

FIGURE 1.17
ERDAS IMAGINE Sentinel-2 MSI band selection and layer output options for layer stack operation. (From ERDAS IMAGINE®/Hexagon Geospatial.)

  f. Click on "**Close**" on the Process List dialog window that appears once it is completed (Figure 1.18).

FIGURE 1.18
Process List for Layer Selection and Stacking operation. (From ERDAS IMAGINE®/Hexagon Geospatial.)

  g. Display the multi-stack image: In ERDAS IMAGINE select: "**File | Open | Raster Layer.**"

h. Navigate to the newly created image: **T31UDQ_20170526T105031_B02-4_8.img**. Remember that the new multi-band image is now an **.img* file. This is ERDAS IMAGINE's default raster data format.

NOTE: Remember in the Select Layer to Add dialog window to change the output file type to "Files of type: **IMAGINE Image (*.img)**" at the drop-down list if the new multi-band image is not available in the directory) (Figure 1.19).

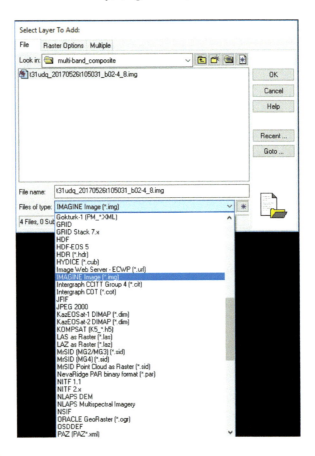

FIGURE 1.19
ERDAS IMAGINE import for IMAGINE Image (*.img) file format. (From ERDAS IMAGINE®/Hexagon Geospatial.)

i. Select the newly created file in the **Select Layer to Add** dialog window and click on the "**Raster Options**" tab. In the "**Layers to Colors**" select Red: **3**, Green: **2**, and Blue: **1**. Next check "**Fit to Frame**" (Figure 1.20).

Acquiring Data: EarthExplorer, GloVis, LandsatLook Viewer   23

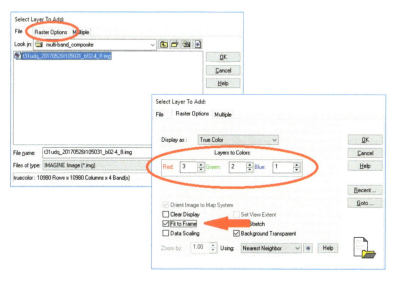

FIGURE 1.20
ERDAS IMAGINE "Layers to Colors" raster options. (From ERDAS IMAGINE®/Hexagon Geospatial.)

j. Select "**OK**" to dismiss the **Select Layer to Add** dialog window and display the multi-band image in the 2D Viewer (Figure 1.21).

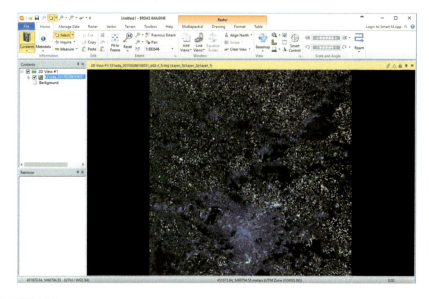

FIGURE 1.21
Completed layer stack result of image T31UDQ_20170526T105031_B02-4_8.img from the ESA's Sentinel-2 MSI dataset of the Paris, France region (Acquisition Date: 2017/05/26). (From ERDAS IMAGINE®/Hexagon Geospatial.)

10. Creating a Layer Stacked image (Landsat 8 data)
    a. First, close any images remaining in the 2D Viewer (**File | Close | Close All Layers**).
    b. Next, from the file menu in ERDAS IMAGINE, click on the "**Raster**" tab. Then select "**Spectral**" from the Resolution grouping. Finally, click on "**Layer Stack**" (Figure 1.22).

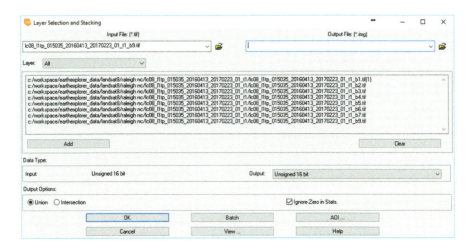

FIGURE 1.22
ERDAS IMAGINE Layer Selection and Stacking dialog window for Landsat 8 OLI/TIRS dataset. (From ERDAS IMAGINE®/Hexagon Geospatial.)

    c. For layer stacking Landsat 8 data (i.e., LC08_L1TP_015035_20160413_20170223_01_T1): Create a layer stacked, or multi-band image, composed of Landsat 8's spectral bands (Band 1–7, and 9) as follows:

    LC08_L1TP_015035_20160413_20170223_01_T1_B1.TIF: Layer: 1
    LC08_L1TP_015035_20160413_20170223_01_T1_B2.TIF: Layer: 1
    LC08_L1TP_015035_20160413_20170223_01_T1_B3.TIF: Layer: 1
    LC08_L1TP_015035_20160413_20170223_01_T1_B4.TIF: Layer: 1
    LC08_L1TP_015035_20160413_20170223_01_T1_B5.TIF: Layer: 1
    LC08_L1TP_015035_20160413_20170223_01_T1_B6.TIF: Layer: 1
    LC08_L1TP_015035_20160413_20170223_01_T1_B7.TIF: Layer: 1
    LC08_L1TP_015035_20160413_20170223_01_T1_B9.TIF: Layer: 1

    It is not necessary to add the following files to the Layer Stack image for this exercise, as these files represent either panchromatic data (Band 8) or thermal infrared data (Bands 10 and 11), as well these bands are captured at different spatial resolutions:

Acquiring Data: EarthExplorer, GloVis, LandsatLook Viewer

LC08_L1TP_015035_20160413_20170223_01_T1_B8.TIF (Panchromatic)

LC08_L1TP_015035_20160413_20170223_01_T1_B10.TIF (Thermal Infrared 1)

LC08_L1TP_015035_20160413_20170223_01_T1_B11.TIF (Thermal Infrared 2)

The new multi-band image will consist of bands 1–7 and band 9 (the original band 9 will be called band 8 in the new multi-band image). Remember that Landsat 8's band 9 will be represented in the multi-band image as band 8, since Landsat 8's band 8 was not included in the layer stack operation.

d. Next, **name** the output file: LC08_L1TP_015035_20160413_201702 23_01_T1_ALB and change the output file type to "Files of type: **IMAGINE Image (*.img)**" from the drop-down list. Note the "ALB" suffix represents a designation that the newly created file will now be comprised of "All Bands" (Bands 1–7, and 9) (Figure 1.23).

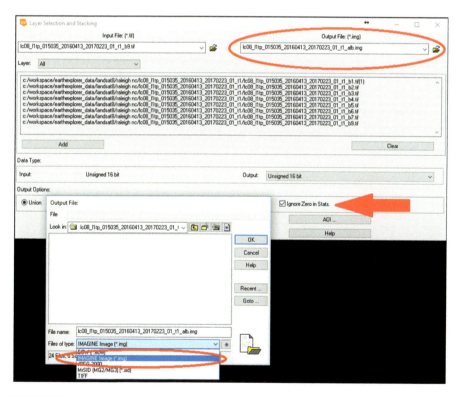

FIGURE 1.23
ERDAS IMAGINE Landsat 8 OLI/TIRS band selection and layer output options for layer stack operation. (From ERDAS IMAGINE®/Hexagon Geospatial.)

The output file will be saved as an *.img file. This is ERDAS IMAGINE's native format for raster files and often the default format that the software will save files in after operations that require creating a new file.

e. Select **"OK"** in the Output File and Layer Selection and Stacking dialog windows.

f. Click on **"Close"** on the Process List dialog window that appears once it is completed.

g. Display the multi-stack image: In ERDAS IMAGINE select: **"File | Open | Raster Layer."**

h. Navigate to the newly created image: LC08_L1TP_015035_20160 413_20170223_01_T1_ALB. Remember that the new multi-band image is now an *.img file. This is ERDAS IMAGINE's default raster data format.

NOTE: Remember in the Select Layer to Add dialog window to change the output file type to "Files of type: **IMAGINE Image (*.img)**" at the drop-down list if you do not see the new multi-band image in the directory) (Figure 1.24).

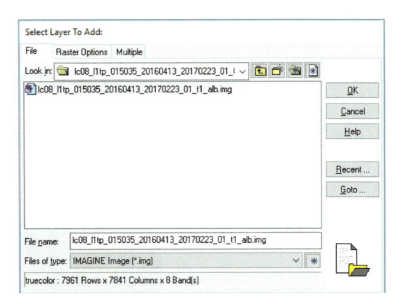

FIGURE 1.24
ERDAS IMAGINE select IMAGINE Image (*.img) layer to add dialog window. (From ERDAS IMAGINE®/Hexagon Geospatial.)

Acquiring Data: EarthExplorer, GloVis, LandsatLook Viewer

i. Select the newly created file in the Select Layer To Add dialog window and on the **"Raster Options"** tab. In the **"Layers to Colors"** select Red: **4**, Green: **3**, and Blue: **2**. Next check **"Fit to Frame"** and **"OK"** to dismiss the Select Layer To Add dialog window and display the multi-band image in the 2D Viewer (Figure 1.25).

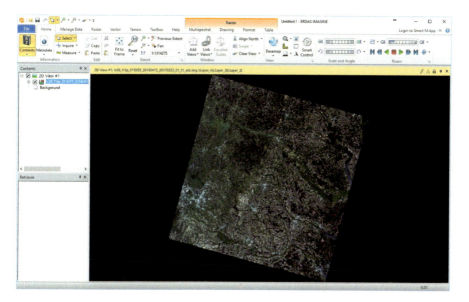

FIGURE 1.25
Completed layer stack result of image LC08_L1TP_015035_20160413_20170223_01_T1_ALB.img from the Landsat 8 OLI/TIRS dataset of the Raleigh, North Carolina region (Acquisition Date: 2016/04/13). (From ERDAS IMAGINE®/Hexagon Geospatial.)

II. Automated Method of Importing EarthExplorer and Creating a Multi-Band, Layer Stack, Image in ERDAS IMAGINE

ERDAS IMAGINE provides an automated method of data import which is particularly useful when downloading data from the USGS EarthExplorer website. However, the newer EarthExplorer naming convention, that represents the USGS Landsat "Collection 1 Level-1" data products, will not currently work in ERDAS IMAGINE's automated data import tool (ERDAS IMAGINE 2016, version 16, build 650). This Landsat naming convention change was effective as of April 30, 2017, on Landsat 7 ETM+ and Landsat 8 OLI/TIRS images. The USGS "Pre-Collection Level-1" inventory, containing Landsat 4–5 TM, Landsat 7 ETM+, and Landsat 8 OLI/TIRS, still using the older naming convention is expected to be no longer be available after

October 1, 2017. Note that the older naming convention (i.e., USGS Landsat Scene ID: LC80150352016104LGN01) must be used as an input until this functionality is updated in ERDAS IMAGINE. However, a workaround is presented here.

1. Select the **Management Data** tab on the Tabbed Ribbons menu and click on the **Import Data** icon. In the **Import** dialog window that opens, select the *"Landsat-7 or Landsat-8 from USGS"* from the format drop-down list (Figure 1.26).

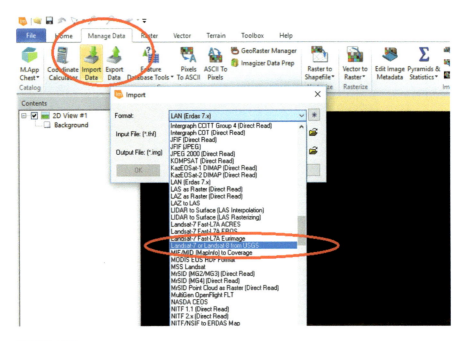

FIGURE 1.26
Import Data option for the "Landsat-7 or Landsat-8 from USGS" format. (From ERDAS IMAGINE®/Hexagon Geospatial.)

Acquiring Data: EarthExplorer, GloVis, LandsatLook Viewer 29

2. In the Import dialog window that opens, select the **Open Directory** icon for the **Input File** list and navigate to where you downloaded the USGS EarthExpolorer unzipped dataset for the Landsat 8 OLI/TIRS imagery of the Raleigh, North Carolina region (File Name: LC08_L1TP_015035_20160413_20170223_01_T1.tar.gz, Acquisition Date: 2016/04/13) (Figure 1.27).

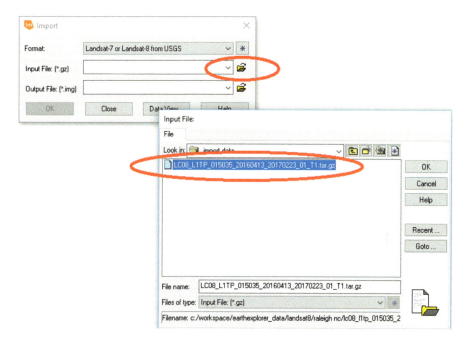

FIGURE 1.27
Data import for the Landsat 8 OLI/TIRS imagery of the Raleigh, North Carolina region (File Name: LC08_L1TP_015035_20160413_20170223_01_T1.tar.gz, Acquisition Date: 2016/04/13). (From ERDAS IMAGINE®/Hexagon Geospatial.)

3. Notice in the Import dialog window that the Output File list box will automatically populate with file name lc08_l1tp_015035_20160413_20170223_01_t1.img. Select OK and an error message appears stating that input file name is not in the original USGS naming convention. Click OK to dismiss this error message (Figure 1.28).

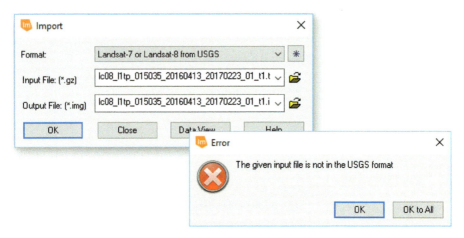

FIGURE 1.28
Error message due to the newer USGS Landsat Collection 1 Level-1 file name (Input File) that is not in the original USGS naming convention. (From ERDAS IMAGINE®/Hexagon Geospatial.)

4. Next, navigate in Windows Explorer to the directory where you downloaded the USGS EarthExpolorer unzipped dataset (LC08_L1TP_015035_20160413_20170223_01_T1.tar.gz). Next, unzip the dataset **twice**! The first unzip operation will unzip the original GZip (*.gz) file to a new file with a TAR (*.tar) extension. In a second unzip operation, unzip the TAR (*tar) file to reveal the multiple image band files in the GeoTIFF (.tif) formats as well as the metadata file, in a text file format (i.e., LC08_L1TP_015035_20160413_20170223_01_T1_MTL.txt).

5. Open the metadata text file and within the notepad document that opens, locate the USGS Landsat Scene ID (e.g., LC80150352016104LGN01). Use the Landsat Scene ID file name to rename the original USGS EarthExplorer unzipped dataset in the Windows Explorer directory (e.g., rename LC08_L1TP_015035_20160413_20170223_01_T1.tar.gz to LC80150352016104LGN01.tar.gz). You can close the metadata text file document and the Windows Explorer directory if you wish (Figure 1.29).

Acquiring Data: EarthExplorer, GloVis, LandsatLook Viewer 31

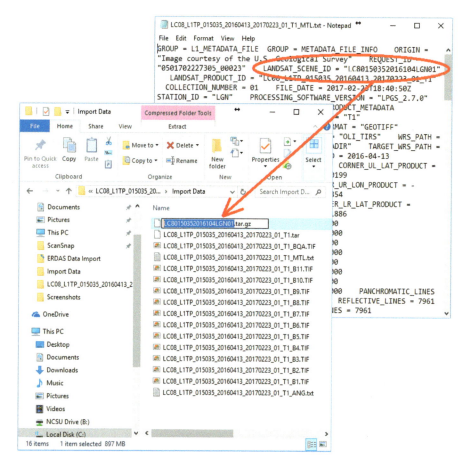

FIGURE 1.29
Rename the original unzipped USGS EarthExpolorer dataset, downloaded for the Landsat 8 OLI/TIRS image (LC08_L1TP_015035_20160413_20170223_01_T1.tar.gz) in Windows Explorer to the Landsat Scene ID (LC80150352016104LGN01.tar.gz) to conform to the original naming convention.

6. In ERDAS IMAGINE, select the **Open Directory** icon the for the Input File list in the Manage Data-Import dialog window. Add the renamed, unzipped, USGS EarthExplorer file name as the **Input File** in the Input File dialog window and select OK. The lc80150352016104lgn01.img file name will be added to the Output File list box in the Import dialog window. Select OK in this window (Figure 1.30).

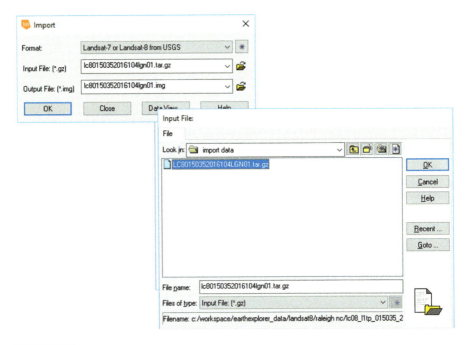

FIGURE 1.30
Add the renamed, unzipped, USGS EarthExplorer file (LC80150352016104LGN01.tar.gz) as the Input File in the Input File dialog window. (From ERDAS IMAGINE®/Hexagon Geospatial.)

7. In the "Import Landsat-7 or Landsat-8 from USGS" dialog window that opens, ensure that a checkmark is placed in front of "**Import Multispectral Data**" (file name: lc80150352016104lgn01-msi.img), and remove all of the other check marks. Next click OK (Figure 1.31).

Acquiring Data: EarthExplorer, GloVis, LandsatLook Viewer 33

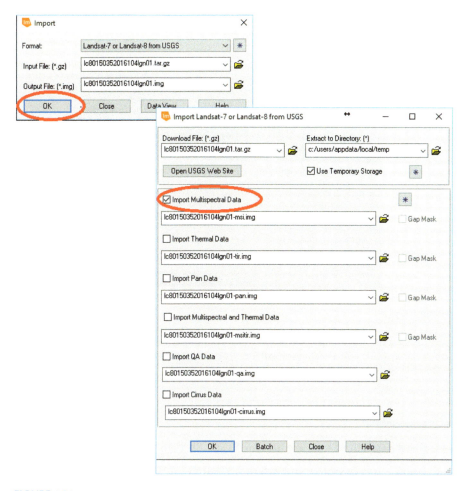

FIGURE 1.31
Import Landsat-7 or Landsat-8 from USGS dialog window. (From ERDAS IMAGINE®/Hexagon Geospatial.)

8. A processing window should open and begin the unzipping and the layer stack process. Once this process completes, the result will be a multi-band image that includes bands 1–7; Band 1 (Coastal aerosol-30 m), Band 2 (Blue-30 m), Band 3 (Green-30 m), Band 4 (Red-30 m), Band 5 (Near Infrared-30 m), Band 6 (Shortwave Infrared 1-30 m), and Band 7 (Shortwave Infrared 2-30 m). Click on close once the Process window is done, and finally click on "Close" on the remaining Import dialog window.

9. Finally, add the new multi-band image (lc80150352016104lgn01-msi.img) using the **File | Open | Raster Layer**... menu option (Figure 1.32).

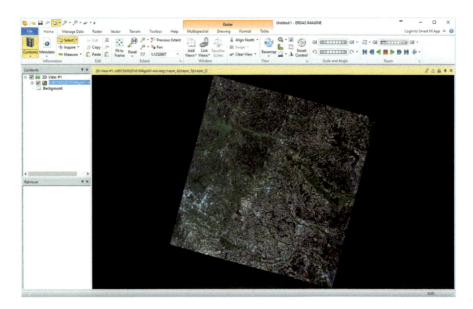

FIGURE 1.32
Completed layer stack result of automated layer stack image downloaded from USGS EarthExplorer for the Landsat 8 OLI/TIRS imagery of the Raleigh, North Carolina region (File Name: LC08_L1TP_015035_20160413_20170223_01_T1.tar.gz, Acquisition Date: 2016/04/13). (From ERDAS IMAGINE®/Hexagon Geospatial.)

III. Finding and Downloading Data in GloVis

An additional option for, perhaps, a faster way to find and download raster data is through the USGS Global Visualization Viewer or **GloVis**. This exercise will use GloVis to acquire a Landsat 8 OLI/TIRS image (USGS Landsat Scene ID: LC80150352016104LGN00, Acquisition Date: 2016/04/13) of Raleigh, North Carolina. Note that the LC80150352016104LGN00 image is the same Landsat 8 OLI/TIRS image previously downloaded from EarthExplorer as LC08_L1TP_015035_20160413_20170223_01_T1 (USGS Landsat Product ID). The new name represents a newer naming convention that is now available in EarthExplorer to represent the USGS Landsat "Collection 1 Level-1" data products. This change was effective as of April 30, 2017, on newly-acquired

Acquiring Data: EarthExplorer, GloVis, LandsatLook Viewer 35

Landsat 7 ETM+ and Landsat 8 OLI/TIRS images. The USGS "Pre-Collection Level-1" inventory, containing Landsat 4-5 TM, Landsat 7 ETM+, and Landsat 8 OLI/TIRS with older naming convention is expected to be no longer available after October 1, 2017.

1. In a web browser, go to the GloVis site (http://glovis.usgs.gov/).
2. Under Dataset in the GloVis browser menu, select: "Landsat Archive | Landsat 8 OLI" to display available Landsat 8 data (Figure 1.33).

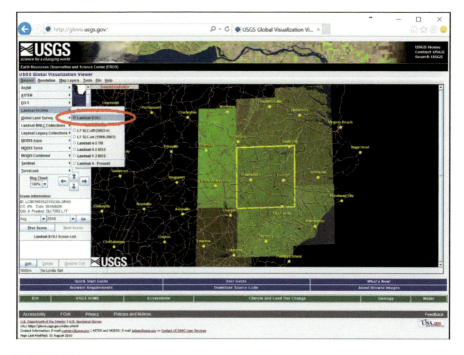

FIGURE 1.33
USGS GloVis web browser Landsat Archive.

NOTE: Notice that other available data collections present in the USGS EarthExplorer browser are here as well. For information about these data formats, see the GloVis online help documentation or the USGS Long Term Archive website: https://lta.cr.usgs.gov/products_overview.

3. Locating data. There are many ways to locate the data of interest in the GloVis Viewer. The scene's Path/Row number, the exact Latitude and Longitude coordinates, or even click on the location on the overview map can be entered.

For this exercise, click on the map location for Raleigh, North Carolina in the viewer's overview map as shown in Figure 1.34.

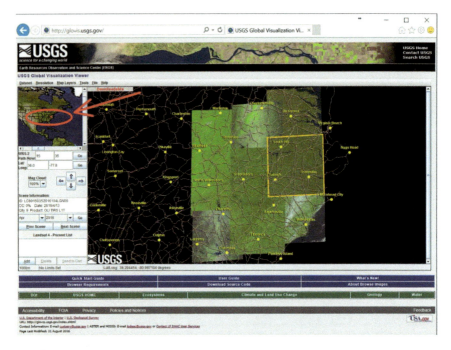

FIGURE 1.34
Location of imagery in the GloVis Viewer.

4. Use the **Max Cloud** option to specify imagery with less than 10% cloud cover. This should provide a generous selection of available images, while reducing the chances of the available images being obscured by clouds for the target area (Figure 1.35).

Acquiring Data: EarthExplorer, GloVis, LandsatLook Viewer 37

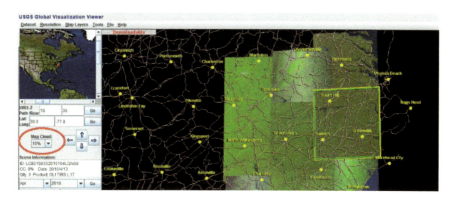

FIGURE 1.35
Selection of percent cloud cover in the GloVis Viewer.

5. Select image LC80150352016104LGN00, Date: 2016/04/13 by clicking "**Add,**" and then "**Send to Cart**" (Figure 1.36).

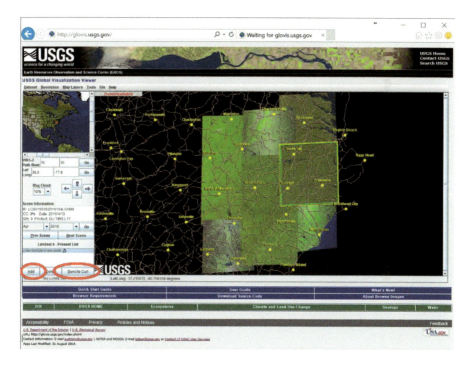

FIGURE 1.36
GloVis add data selection.

38 *Image Processing and Data Analysis with ERDAS IMAGINE®*

6. Review the metadata that is next displayed to determine recorded image quality and the actual amount of cloud cover in the scene so as not to obscure the area of interest (Figure 1.37).

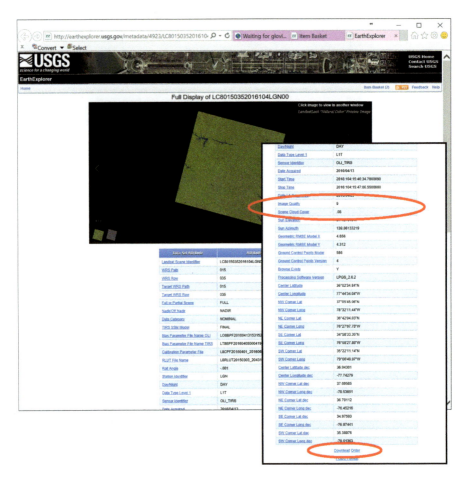

FIGURE 1.37
GloVis review of metadata option.

Acquiring Data: EarthExplorer, GloVis, LandsatLook Viewer 39

7. Finally, select "**Download**" to begin the download process of the file (Figure 1.38).

FIGURE 1.38
GloVis data download option.

Follow the directions in the previous section (I. Finding and downloading data in EarthExplorer), beginning with "**Step 7. Unzipping the image data**" to unzip and get the data into ERDAS IMAGINE.

IV. Displaying Raster Data and Creating a Multi-Band Image in Esri ArcMap ArcGIS for Desktop

Environmental Systems Research Institute's (Esri) ArcGIS for Desktop or ArcMap is currently the industry-leading geographic information system (GIS) application package for vector, raster, and relational database management. This software provides a suite of tools that allows the user to create and work with maps, organize and store geographic data, and perform geospatial analyses. While ArcMap does possess very powerful digital image processing features, the software continues to develop its capabilities in comparison to dedicated packages, such as ERDAS IMAGINE. Esri currently licenses ArcMap Desktop on the Microsoft Windows platform in three versions. The first version is ArcGIS for Desktop Basic (previously known as ArcView), which provides basic spatial analysis functions (buffer, clip, overlay, etc.), general spatial analysis, view and combine spatial data layers, and output created maps. The intermediate version is ArcGIS for Desktop Standard (previously known as ArcEditor), which includes all the features of the ArcGIS for Desktop Basic, as well as incorporating tools targeting the advanced editing of vector and geodatabase features. The most comprehensive version is ArcGIS for Desktop Advanced (previously known as ArcInfo), which includes advanced geospatial spatial editing, analysis, and modeling tools.

This exercise will display in ArcGIS and create a multi-band image of the Sentinel-2 MSI of the Paris, France region (File Name: L1C_T31UDQ_A010058_20170526T105518, Acquisition Date: 2017/05/26) previously downloaded from EarthExplorer in the previous section.

1. From Windows program or Start Menu, launch the ArcGIS ArcMap application. If the "ArcMap—Getting Started" dialog box appears, simply click on "Cancel" to dismiss this window (Figure 1.39).

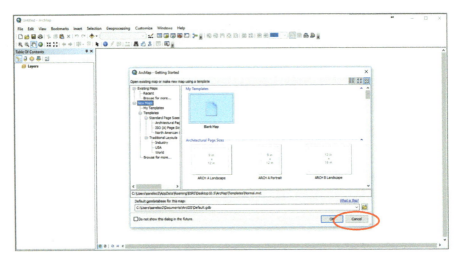

FIGURE 1.39
Esri's ArcMap ArcGIS for Desktop application interface. (From ARCMap/ESRI.)

2. From the upper file menu of icons, click on the **Add Data** icon. In the Add Data dialogue that opens, click on the **Connect to Folder** icon and navigate to the directory in which you downloaded and unzipped the Sentinel-2 imagery data for Paris, France and click on "**OK**" (Figure 1.40).

Acquiring Data: EarthExplorer, GloVis, LandsatLook Viewer 41

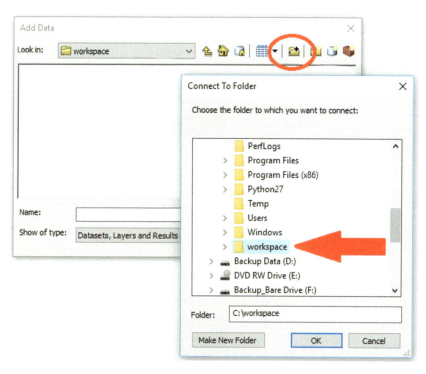

FIGURE 1.40
ArcMap connect to folder option. (From ARCMap/ESRI.)

3. Navigate to the individual JPEG 2000 files for the Paris, France data (…S2A_MSIL1C_20170526T105031_N0205_R051_T31UDQ_ 20170526T105518.SAFE\GRANULE\L1C_T31UDQ_A010058_ 20170526T105518\IMG_DATA), and one by one add the JPEG 2000 files for bands 2, 3, 4, and 8 (i.e., T31UDQ_20170526T105031_B02.jp2, T31UDQ_20170526T105031_B03.jp2, T31UDQ_20170526T105031_ B04.jp2, and T31UDQ_20170526T105031_B08.jp2) to the ArcMap map document window (Figure 1.41).

42   *Image Processing and Data Analysis with ERDAS IMAGINE®*

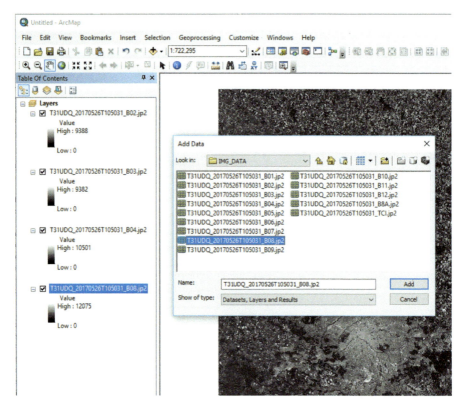

FIGURE 1.41
ArcMap add data option. (From ARCMap/ESRI.)

4. Rearrange each file in the Table of Contents (file layers along the left side of the ArcMap interface) so that bands arranged in order with Band 2 at the top, and Band 8 at the bottom.

5. From the Windows file menu, choose Image Analysis. Next highlight all four image bands (Ctrl-Click) at the top of the Image Analysis window, and then go the "Processing" section and click on the **Composite Bands** icon . This will create a temporary multi-band image composite (Figure 1.42).

6. Single-click on the new layer added to the top of the Table of Contents to make the name editable and rename it "Composite_ T31UDQ_20170526T105031_B02-4_8.jp2" to denote that the new composite image created in ArcMap will now be comprised of the original Sentinel-2 bands 2–4, and 8.

Acquiring Data: EarthExplorer, GloVis, LandsatLook Viewer 43

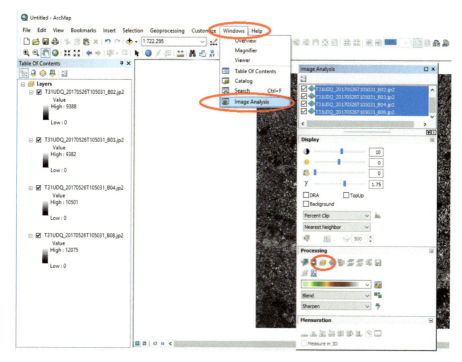

FIGURE 1.42
ArcMap Image Analyst option. (From ARCMap/ESRI.)

7. Next, right-click on the "Composite_T31UDQ_20170526T105031_B02-4_8.jp2" layer in the Table of Contents and, in the pop-up list, select Data | Export Data. This will open an "Export Raster Data" window. Navigate to the directory where you wish to save the new file and click on "Save." In the Output Raster window that pops up, click on "No" to prevent ArcMap from adding NODATA pixels to the new file. Finally, add this new file to the map document by selecting "Yes" when presented with the option (Figure 1.43).

8. Rename the new file "ArcMap_T31UDQ_20170526T105031_B02-4_8.tif" to designate that it was Layer Stacked in ArcGIS ArcMap.

44 *Image Processing and Data Analysis with ERDAS IMAGINE®*

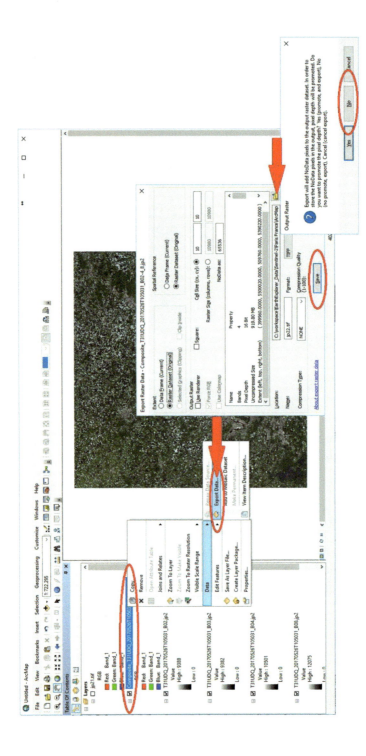

FIGURE 1.43
ArcMap Export Raster Data option. (From ARCMap/ESRI.)

Acquiring Data: EarthExplorer, GloVis, LandsatLook Viewer **45**

V. Displaying Raster Data and Creating a Multi-Band Image in Quantum Geographic Information Systems

Quantum GIS (QGIS) is an open source package that possesses many geospatial processing features that are used for processing both vector and raster data, as well as developing sophisticated spatial analysis models. The uniqueness of this application package is that it is freely available, and the functionality continues to expand.

Once downloaded and installed, the QGIS the application can be started from the Windows Programs/Start menu, and then selecting QGIS Desktop. This exercise will display in ArcGIS and create a multi-band image of the Sentinel-2 MSI of the Paris, France region (File Name: L1C_T31UDQ_A010058_20170526T105518, Acquisition Date: 2017/05/26) previously downloaded from EarthExplorer in the earlier section.

1. To create a multi-band image in QGIS, from the file menu at the top of the QGIS interface, select: **Raster | Miscellaneous | Merge...**

2. In the Merge dialog window that opens, for Input files choose the select button and navigate to the individual JPEG 2000 files for the Paris France data (...S2A_MSIL1C_20170526T105031_N0205_R051_T31UDQ_20170526T105518.SAFE\GRANULE\L1C_T31UDQ_A010058_20170526T105518\IMG_DATA). Hold down the Ctrl key on your computer keyboard and select the JPEG 2000 files for bands 2, 3, 4, and 8 (e.g., T31UDQ_20170526T105031_B02.jp2, T31UDQ_20170526T105031_B03.jp2, T31UDQ_20170526T105031_B04.jp2, and T31UDQ_20170526T105031_B08.jp2) and click on "**Open**" in the Select the file to Merge window (Figure 1.44).

46   *Image Processing and Data Analysis with ERDAS IMAGINE®*

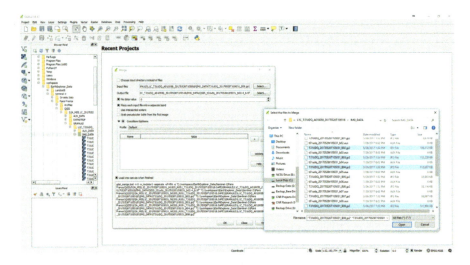

FIGURE 1.44
Quantum GIS (QGIS) application interface. (From QGIS.)

3. In the Output file field, click on the Select button to navigate to where you would like to save the file, and name this file "QGIS_T31UDQ_20170526T105031_B02-4_8.tif" to denote that the new composite image created in QGIS will now be comprised of the original Sentinel-2 bands 2–4, and 8.
4. Next place checkmarks beside "**No date of value**" and "**Place each input file into a separate band.**"
5. Finally, click on "**OK**" to begin the merge process and add the multi-band composite image to the display view in QGIS (Figure 1.45).

Acquiring Data: EarthExplorer, GloVis, LandsatLook Viewer

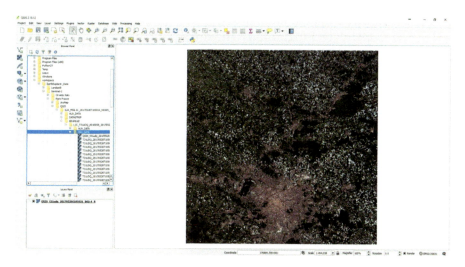

FIGURE 1.45
Completed layer stack (band merge) in QGIS. (From QGIS.)

Review Questions

1. Remote sensing can be characterized as being the science of acquiring any information about an object or a phenomenon on the surface of the Earth without coming into physical contact with it. In what data form is this information typically acquired in?

2. Data captured from remote sensing of earth features are also typically based on four types of resolutions. Name these resolutions.

3. Electromagnetic radiation (EMR) is defined as all energy that moves with the velocity of light in a harmonic wave pattern. How could EMR be explained further?

4. The Landsat 8 Operational Land Imager (OLI) and Thermal Infrared Sensor (TIRS) or simply Landsat 8, was launched on February 11, 2013, and represents the United States' continuing mission of the Landsat sensor program. Within how many spectral bands are these images captured? What is the spatial resolution of each spectral band? How many panchromatic bands are included in this sensor and at what spatial resolution? What are the (resampled) spatial resolutions of the thermal bands? What is the revisit cycle of the Landsat 8 satellite?

5. The Sentinel-2 satellite multispectral sensor is a component of the ESA's Copernicus Programme. The initial Sentinel-2A satellite was launched June 23, 2015. This was followed by a Sentinel-2B satellite launch on March 7, 2017. The two Sentinel satellites (2A and 2B) operate in a 180° phase with the orbit allowing for global coverage of the Earth's surface every five days, or a lower temporal resolution at higher latitudes (every 10 days with one satellite). How many spectral bands are the Sentinel sensors' images available in? What are the spatial resolutions of these the spectral bands?

2

Introduction to Image Data Processing

Overview

Raster imagery is typically analyzed, in a computer desktop environment, with specialized image processing software packages. Although the availability and the functionality of these packages are ever-changing, a great number of packages are currently available (Table 2.1). In general, each of these packages may have strengths and weaknesses in many different areas of image processing, such as radiometric, spatial, or spectral processing applications, as well as vector, spectral/hyper-spectral, or object-based modeling.

The ERDAS IMAGINE® software runs on Microsoft Windows-based platforms and can incorporate the majority of raster and vector data formats. This software is periodically updated; however, the 2016 version is the most updated version that is demonstrated here. The majority of the principles described here remain applicable for future updates as well (Hexagon Geospatial's website for further information: http://www.hexagongeo spatial.com/products/producer-suite/erdas-imagine).

While ERDAS IMAGINE purports to be the industry-leading standard in digital image processing, a host of other digital image processing packages are also available commercially, and even for free. Table 2.1 lists several commonly available image processing packages.

There are a number of digital image analysis packages currently available. Some will be more feature packed and user-adaptable than others. However, any investigator should analyze their current available options as it relates to their own specific needs. Available solutions do not have to be cost prohibitive. Freeware and open source packages are becoming more available and highly competitive to many of the commercial packages. The techniques demonstrated in this textbook, while demonstrated utilizing the Hexagon

49

50   *Image Processing and Data Analysis with ERDAS IMAGINE®*

TABLE 2.1

Table of Commonly Available Image Processing Packages

Commercial Software		
ENVI	Harris Geospatial	http://www.harrisgeospatial.com/ProductsandSolutions/GeospatialProducts/ENVI.aspx
Google Earth Engine (GEE)	Google	https://earthengine.google.com/
ArcMap ArcGIS for Desktop	ESRI	https://www.arcgis.com/features/index.html
ERDAS ER Mapper (formally ER Mapper)	Hexagon Geospatial	www.hexagongeospatial.com/products/power-portfolio/er-mapper
GeoMedia	Hexagon Geospatial	http://www.hexagongeospatial.com/products/power-portfolio/geomedia
Geomatica	PCI	http://www.pcigeomatics.com/
TNTmips	MircoImages	http://www.microimages.com/products/tntmips.htm
TerrSet (formerly IDRISI)	Clark Labs	https://clarklabs.org/
eCognition	Trimble	http://www.ecognition.com/
TerraExplorer	Skyline Software Systems	http://www.skylineglobe.com/SkylineGlobe/corporate/Default.aspx
SOCET SET	BAE Systems Geospatial eXploitation Products (GXP)	http://www.geospatialexploitationproducts.com/content/socet-set-v56/
GeoRover	Leidos	https://www.leidos.com/products/software/georover
Freeware		
MultiSpec	Purdue University	https://engineering.purdue.edu/~biehl/MultiSpec/
PCRaster	Utrecht University, Faculty of Geosciences	http://pcraster.geo.uu.nl/
Sentinel Application Platform (SNAP)	European Space Agency (ESA)	http://step.esa.int/main/toolboxes/snap/
BILKO	European Space Agency (ESA) LearnEO! project	http://www.learn-eo.org/software.php
Integrated Land and Water Information System (ILWIS)	World Institute for Conservation and Environment (WICE)	http://www.ilwis.org/
FUSION	University of Washington and USDA Forest Service	http://forsys.cfr.washington.edu/fusion/fusionlatest.html
Elshayal Smart GIS	First African Arabian Egyptian GIS GPS Software by Smart GIS (Mohamed Elshayal)	http://freesmartgis.blogspot.com/

(Continued)

Introduction to Image Data Processing

TABLE 2.1 (*Continued*)

Table of Commonly Available Image Processing Packages

Commercial Software

Open Source

GRASS GIS	GRASS GIS project	http://grass.osgeo.org/
Quantum GIS (QGIS)	QGIS volunteers	http://www.qgis.org/en/site/
SAGA	SAGA User Group and J. Böhner and O. Conrad of the University of Hamburg, Germany	http://saga-gis.org/en/index.html
BEAM	Brockmann Consultants	http://www.brockmann-consult.de/cms/web/beam/
Spring Project	Brazil National Institute for Space Research	http://www.dpi.inpe.br/spring/english/index.html
OpenEV	OpenEV Developers	http://openev.sourceforge.net/
Opticks	Optics Group with support from Ball Aerospace and the U.S. Air Force	https://opticks.org/
Orfeo toolbox (OTB—ORFEO Toolbox Library)	ORFEO Accompaniment Program with support from the Centre National D'etudes Spatiales (CNES), France	https://www.orfeo-toolbox.org/
OSSIM—Open Source Software Image Map Documentation	funded by several US government agencies in the intelligence and defense community	http://trac.osgeo.org/ossim/
R Project for Statistical Computing	The R Foundation	https://www.r-project.org/
SNAP—SeNtinel Application Platform	European Space Agency, Brockmann Consultants, Array Systems Computing and C-S	http://step.esa.int/main/toolboxes/snap/

Geospatial Intelligence

RemoteView	Textron Systems	http://www.textronsystems.com/what-we-do/advanced-information-solutions
SOCET GXP	BAE Systems Geospatial eXploitation Products (GXP)	http://www.geospatialexploitationproducts.com/socet-gxp/
FalconView	Georgia Tech Research Institute for the U.S. Department of Defense	https://www.falconview.org/trac/FalconView
Common Spectral MASINT Exploitation Capability (COSMEC)	unavailable	unavailable

Geospatial's ERDAS Image software, are geared towards being applicable in any software package. One must be aware that this software package goes through a number of updating processes and often has to look for the updates. However, the main functionality of this package remains relatively applicable to all updated versions. For comparative purposes, certain applications are additionally demonstrated with Esri's ArcMap ArcGIS for Desktop and open source QGIS.

Introduction to Digital Image Processing Application

The objective of this exercise is to give you the opportunity to become familiar with some of the capabilities of this software package and the data that may be used within the software.

As a general framework for working with and storing the data for using in ERDAS IMAGINE, it is advisable to dedicate a working directory that grants you full read and write permissions. Another consideration should be storing the data in a directory that doesn't contain a long directory path with spaces in the path. For example, a good directory structure would be to create a workspace folder directly under **C:** or **C:\users**, or even a workspace that is specifically dedicated to the image analysis work, such as **C:\ workspace** or **C:\users\workspace**. In the following, you will learn to set ERDAS to open this location as the default location. Once the default location has been setup, you may choose to copy all the supplied example data to this directory for faster access.

Learning Objectives

1. To develop a basic familiarization with the digital image processing software package, ERDAS IMAGINE.
2. To explore remotely sensed digital data.

Data required: The data used in this exercise will give you a chance to practice the skills you acquired in the previous chapter at downloading data from the choice of either the USGS EarthExplorer (http://earthexplorer.usgs.gov/)

Introduction to Image Data Processing 53

or USGS GloVis (http://glovis.usgs.gov/) websites. This exercise will acquire both a Landsat 8 and Sentinel-2 image of Rio de Janeiro, Brazil:

1. Landsat 8: LC08_L1TP_217076_20170218_20170228_01_T1 (Acquisition Date: 2017/02/18)
2. Sentinel-2: L1C_T23KPQ_A008672_20170218T130353 (Acquisition Date: 2017/02/18)

I. Obtaining Required Data in EarthExplorer

From the USGS EarthExplorer website, this exercise will download a Landsat 8 image and Sentinel-2 image acquired on the same date (image acquisition date: 2017/02/18) of Rio de Janeiro, Brazil.

1. In the Address/Place field, type in the word *"Rio de Janeiro"* and click on the **"Show"** button below the box. Next, click on the **Rio de Janeiro, State of Rio de Janeiro, Brazil** link under the Address/Place heading to show the location on the map and add coordinates to the Area of Interest Control.
2. In the **"Search from"** boxes in the Search Criteria section enter the following dates: **05/01/2017** to **08/31/2017**. Next select **"Data Sets"** below Date Range options.
3. Select **"Landsat Archive | L8 OLI/TIRS C1 Level-1"** and the **"Sentinel | Sentinel-2"** categories on the Data Sets tab.
4. On the Additional Criteria tab, scroll down to the **Land Cloud Cover** option and select **"Less than 10%."** Also, scroll down to the **Scene Cloud Cover** option and select **"Less than 10%"** as well.
5. Next select **"Results"** at the bottom of Data Sets tab. Now inspect the available Landsat 8 image and Sentinel-2 imagery and click on the **Show Metadata and Browse** icon for both the Landsat 8: **LC08_L1TP_217076_20170218_20170228_01_T1** and Sentinel-2: **L1C_T23KPQ_A008672_20170218T130353** images (Figure 2.1).

FIGURE 2.1
EarthExplorer data search for imagery of Rio de Janeiro, State of Rio de Janeiro, Brazil. On left: Landsat 8 OLI/TIRS (File name: LC08_L1TP_217076_2 0170218_20170228_01_T1; Acquisition date: 2017/02/18), and on right: Sentinel-2 MSI (File name: L1C_T23KPQ_A008672_20170218T130353; Acquisition date: 2017/02/18) images.

Introduction to Image Data Processing 55

6. To download the images, select the **Download Options** icon ⬇ to begin downloading the Landsat 8 and Sentinel-2 products. Continue to follow the directions discussed previously in **Chapter 1** to unzip the image (**step 7. Unzipping the L8 OLI/TIRS data**) and create a multi-band image (**step 10. Creating a Layer Stacked image (Landsat 8 data)**) with the appropriate bands in ERDAS IMAGINE.

   NOTE: For the Landsat 8 image, remember the new multi-band image will consist of bands 1–7 and band 9 (the original band 9 will be called band 8 in the new multi-band image). Landsat 8's band 8 (panchromatic data), and band 10 and 11(thermal infrared data) are captured at different spatial resolutions and should not be included in the layer stack operation. For the Sentinel-2 image, remember the new multi-band image will consist of bands 2–4 and band 8. The original bands 2, 3, and 4 will now be bands 1, 2, and 3 in the new multi-band image. Also, the original band 8 will be called band 4 in the new multi-band image. Sentinel-2's band 1, and bands 5–7, 8A, and 9–12 are captured at different spatial resolutions and should not be included in the layer stack operation.

7. Name the Landsat 8 output file: **LC08_L1TP_217076_20170218_ 20170228_01_T1_ALB**, and Sentinel-2: **T23KPQ_A008672_ 20170218T130353_B02-4_8**. Be sure to change the output file type to "Files of type: **IMAGINE Image (*.img)**" at the drop-down list for each layer stack file. Note the "ALB" suffix for the Landsat 8 image represents a designation that the newly created file will now be comprised of "All Bands" (e.g., Bands 1–7, and 9). The "**B02–4_8**" suffix for the Sentinel-2 image represents a designation that the newly created file will now be comprised of Bands 2–4, and 8).

8. Also, be sure to check "**Ignore Zero in Stats.**" Finally, click on "**OK**" to begin the layer stack conversion and dismiss the Process List once the Layer Stack is complete.

The Layer Stack band selections should resemble Figure 2.2.

56   *Image Processing and Data Analysis with ERDAS IMAGINE®*

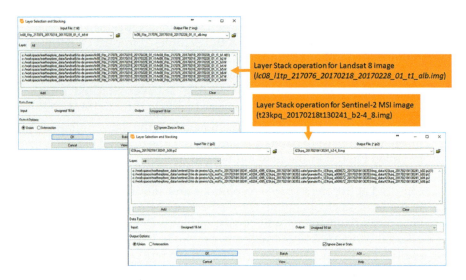

FIGURE 2.2
ERDAS IMAGINE Layer Selection and Stacking operation. (From ERDAS IMAGINE®/Hexagon Geospatial.)

II. ERDAS IMAGINE Graphic User Interface

Understanding how the geographic user interface (GUI) within the ERDAS IMAGINE software is laid out is the key to successfully accessing and analyzing data within the software package. Components of this exercise adapted from University of Wyoming on-line resources, created by Dr. Ken Driese and available at: http://www.uwyo.edu/rs4111/.

When ERDAS IMAGINE is started (Start/All Programs /ERDAS IMAGINE /ERDAS IMAGINE) the main software interface (Graphic User Interface or GUI) window opens and includes three major components:

1. **Tabbed Ribbons** (menu icons across the top)
2. "**Contents**" and "**Retriever**" legend boxes (below the Tabbed Ribbons and to the left of the GUI)
   a. The **Contents** box lists all the files that are displayed in the graphic window.
   b. The **Retriever** box stores links that may be added for quick access, similar to the process of creating a bookmarked link in an internet web browser. You can drag image icons from the Contents to the Retriever box (or *vice versa*), and you can save Retriever box files **as *.ipx files** and reopen them later.
3. **Graphic View Window** (to the right of the Contents and Retriever legend boxes) (Figure 2.3).

Introduction to Image Data Processing

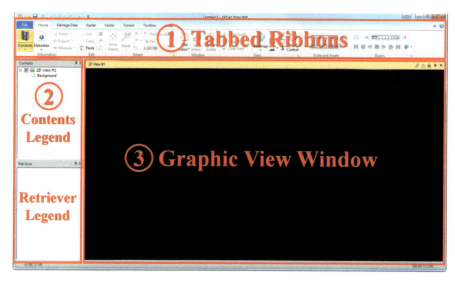

FIGURE 2.3
ERDAS IMAGINE interface layout. (From ERDAS IMAGINE®/Hexagon Geospatial.)

Exploring ERDAS Help Documents

The first step in becoming familiar with ERDAS IMAGINE is to explore the help documents. Multiple ERDAS help options are accessible from the question mark icon [?] found on the top right corner of the ERDAS interface. (Figure 2.4).

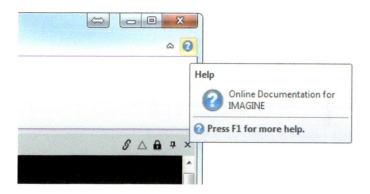

FIGURE 2.4
ERDAS IMAGINE Help icon. (From ERDAS IMAGINE®/Hexagon Geospatial.)

Selecting the Help icon opens a separate internet browser containing a contents category list for selecting various operations within the software.

To explore the help documentation, select the following:

Help icon 🌀 | **IMAGINE Interface** | **IMAGINE Workspace**

Once the "**IMAGINE Ribbon Workspace—Getting Started**" page opens in the browser, spend some time getting familiar with the ERDAS IMAGINE interface (Figure 2.5).

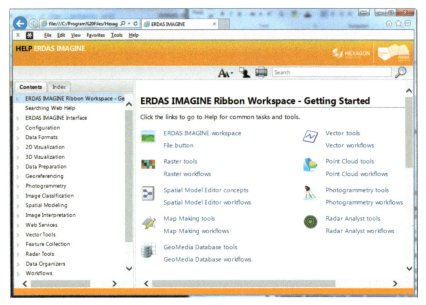

FIGURE 2.5
ERDAS IMAGINE Online Help documentation. (From ERDAS IMAGINE®/Hexagon Geospatial.)

Online Help Documents: IMAGINE Ribbon Workspace – Getting Started

Help documents for additional tools within the software are accessible from the menu bar's **Help** tab. You may access these documents by clicking on the menu bar's **Help** tab on the main interface.

- **Search Library** provides the ability perform a search of ERDAS IMAGINE documents by keywords.
- **Field Guides** provide both theoretical and conceptual explanation of some of the functions in ERDAS.
- **Tour Guides** provide tutorial guides to some of the more commonly used functions in ERDAS.
- **User Guides** provide tutorials on a number of ERDAS applications.
- **Language Reference** provides a reference of the syntax programming necessary to create custom models in ERDAS.

Introduction to Image Data Processing   59

Setting Up Workspace Preferences

The **Preferences Editor** will allow you to specify the default data storage and retrieval (workspace) locations that ERDAS IMAGINE will use for your data.
Workspace Preferences:

1. Click on the menu bar's **File Tab**. In the dialog box that opens, click on the "**Preferences**" button at the bottom (Figure 2.6). In the Preference Editor dialog box that opens, the "**User Interface & Session**" category will be selected by default.

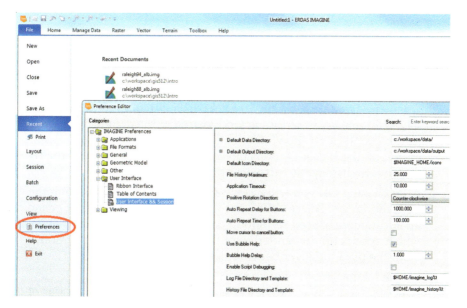

FIGURE 2.6
ERDAS IMAGINE workspace preferences setup. (From ERDAS IMAGINE®/Hexagon Geospatial.)

2. Set the home directory of your workspace location where you copied the example data to:

   Default Data Directory: Change this directory location to the working directory you intend on storing and retrieving your data, e.g.:

   C:/workspace/data

   Default Output Directory: Change this directory location to the working directory you specified above, e.g.:

   C:/workspace/data/output

   Click on "**User Save**" to save your preferences.

Opening Images

Data imagery in ERDAS IMAGINE can be loaded in several ways. The current data you downloaded is in a native ERDAS ***.img** format so at this point no additional importation steps are required.

Open an image:

1. On the **Home Tab** (which is active by default when you start ERDAS), note that many options on the ribbon are inactive. These options will activate once data is loaded into the viewer.

Open a Landsat Thematic Mapper (TM) image:

2. Select the **Open Folder** icon [icon] in the menu bar at the top of the ERDAS IMAGINE interface. If preferences were set correctly, you should see the image files that you copied into your default directory from the companion media.
3. Open the **Landsat 8 (LC08_L1TP_217076_20170218_20170228_01_T1_ALB)** image of the Rio de Janeiro area in the default 2D Viewer (black graphics window). Remember to select the "Fit to Frame" option on the Raster Options tab in the **Select Layer to Add** dialog window to display the full extent of the data.
4. After clicking **OK** you will receive a message asking you to compute pyramid layers. Select "**YES**" to this option. This option will create an additional reduced resolution data file (***.rrd**), or pyramid layer, which speeds up the rendering of the image in the viewer window. Once the pyramid layer has completed, click on **close**.

   The image should now be open in the **2D View #1** graphic view window. Note when the viewer is active, the top viewer header of the frame will turn a yellowish-orange color.

   Now open an image in a second 2D viewer window.
5. From the **Home** tab ribbon: Click on **Add Views/Create New 2D View** (Figure 2.7).

   A second graphics window should open beside the Landsat 8 image. Note that when the new 2D viewer has been added the frame or header of this view will now be active, as signified by the yellow viewer header.

Introduction to Image Data Processing 61

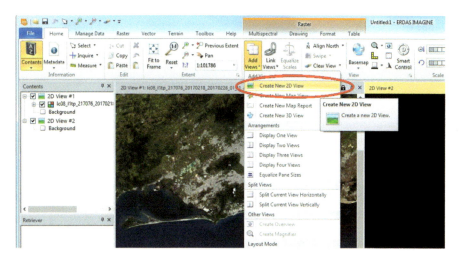

FIGURE 2.7
ERDAS IMAGINE Create New 2D View. (From ERDAS IMAGINE®/Hexagon Geospatial.)

6. Open the **Sentinel-2 (T23KPQ_A008672_20170218T130353_B02-4_8)** image in the **2D View #2** window. Remember to select the "Fit to Frame" option on the Raster Options tab in the Select Layer to Add dialog window to display the full extent of the data. Then click on "OK" (Figure 2.8).

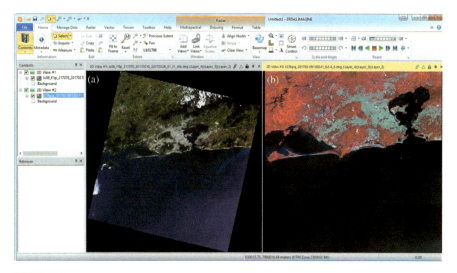

FIGURE 2.8
Landsat 8 OLI/TIRS image (LC08_L1TP_217076_20170218_20170228_01_T1_ALB) of the Rio de Janeiro area displayed in the default 2D Viewer (a) and the Sentinel-2 (T23KPQ_A008672_20170218T130353_B02-4_8) image displayed in a second 2D Viewer (b). (From ERDAS IMAGINE®/Hexagon Geospatial.)

Click on the **Fit to Frame** ⬚ icon to display the full extent of the data in the viewer window. Do this in both viewers by first selecting the viewer to activate it, and then selecting the Fit to Frame option.

Getting Data Information

To get more information about the image files you will need to select the Metadata option from the Home tab.

Metadata Icon:

Click on the **Metadata icon** ⓘ from the Home tab. An "Image Metadata" window should open that includes some basic information about the image in the active viewer.

Review the following information for both the Land 8 and Sentinel-2 images:

1. Number of layers (bands)
2. Number of columns (width)
3. Number of rows (height)
4. Data Type
5. Minimum, maximum and mean pixel values
6. Projection, Spheroid, and Datum

Click close on the *Image Metadata* window to dismiss it.

Band Combinations

In many image processing packages the image displayed is determined by the combinations of satellite image-spectral bands used for the display and the three additive primary colors (red, green and blue) that are associated with those bands. Each image spectral band has captured a different range of wavelengths (i.e. band) within the electromagnetic spectrum. By assigning colors that represent a *True Color* or natural color image, means that the image is displayed in colors that are similar to how your eyes would normally see them (i.e. grass is green and water is blue). A "False Color" image means that the colors have been assigned to three different wavelengths, which your eyes might not normally see (e.g., vegetation appears in shades of red).

A comparison of spectral wavelengths (micrometers) and spatial cell size resolutions (meters) for three of the most recent Landsat sensors are listed in Table 2.2. Landsat 5 TM (Thematic Mapper), Landsat 7 ETM+ (Enhanced Thematic Mapper Plus), and Landsat 8 OLI (Operational Land Imager) and TIRS (Thermal Infrared Sensor) approximate scene sizes are 170 km north-south by 183 km east-west (106 mi by 114 mi) (Figure 2.9).

Introduction to Image Data Processing

TABLE 2.2

Comparison of Landsat 5 TM, 7 ETM+, and 8 OLI/TIRS Sensor Bands and Ground Spatial Resolutions

Bands	Landsat 5 Wavelengths (μm)/Cell Size Resolution (m)	Landsat 7 Wavelengths (μm)/Cell Size Resolution (m)	Landsat 8 Wavelengths (μm)/Cell Size Resolution (m)
Band 1			0.43–0.45/30
Band 1	0.45–0.52/30	0.45-0.52/30	
Band 2	0.52–0.60/30	0.52–0.60/30	0.45–0.51/30
Band 3	0.63–0.69/30	0.63–0.69/30	0.53–0.59/30
Band 4	0.76–0.90/30	0.77–0.90/30	0.64–0.67/30
Band 5	1.55–1.75/30	1.55–1.75/30	0.85–0.88/30
Band 6	10.40–12.50/120[a]	10.4012.50/60[b]	1.57–1.65/30
Band 7	2.08–2.35/30	2.09–2.35/30	2.11–2.29/30
Band 8		0.52–0.90/15	0.50–0.68/15
Band 9			1.36–1.38/30
Band 10			10.60–11.19 100[c]
Band 11			11.50–12.51 100[c]

[a] Landsat 5 TM Band 6 (thermal infrared) was acquired at 120 m but were resampled to a 30 m pixel resolution.
[b] Landsat 7 ETM+ Band 6 (thermal infrared) was acquired at 60 m but were resampled to a 30 m pixel resolution.
[c] Landsat 8 OLS and TIRS Bands 10 and 11 are acquired at 100 m but are resampled to a 30 m pixel resolution. Band 8 is the panchromatic band in the Landsat 7 ETM+ and the Landsat 8 OLS and TIRS sensors.

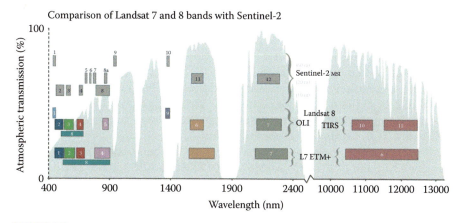

FIGURE 2.9
Comparison of Landsat 7 ETM+ and Landsat 9 OLI/TIRS sensors with Sentinel-2 MSI sensor band placement. (Courtesy of U.S. Geological Survey (https://eros.usgs.gov/sentinel-2).)

64   *Image Processing and Data Analysis with ERDAS IMAGINE®*

The Landsat example data was acquired from the Landsat 8 sensor. When opening this image in ERDAS IMAGINE, the display defaults to a "standard" False Color Composite (FCC) display that is associated with the previous Landsat 5 TM and Landsat 7 ETM+ sensors, using a 4 (red), 3 (green), 2 (blue) band combination. However, Landsat 8 adds a new Band 1 (coastal aerosol) that essentially shifts the traditional band placements used for the previous two Landsat sensors. Thus, the default 4, 3, 2 band combination displays a FCC when opening Landsat 5 TM and Landsat 7 ETM+ images, and a True Color Composite (TCC) when opening a Landsat 8 OLI/TIRS image.

With the Sentinel-2 MSI sensor, when opening the image in ERDAS IMAGINE, the image appears as a FCC band combination. By default, the software assigns the same standard 4 (red), 3 (green), 2 (blue) band combination. However, remember in the previously created Sentinel-2 multi-band image, band 1 was not included in the original layer stack. This means the bands being displayed are actually 8 (red), 4 (green), and 3 (blue). This band combination displays a FCC in the 2D View window. A Sentinel-2 TCC (originally achieved by 4, 3, 2) would be achieved by a creating a 3 (red), 2 (green), 1 (blue) band combination from the multi-band image since band 1 was not included in the layer stack operation.

Create a False Color Composite Display Band Combination

Click on the 2D View #1 graphics viewer window containing the Landsat 8 image of the Rio de Janeiro area (**LC08_L1TP_217076_20170218_20170228_01_T1_ALB**). The frame turns yellow indicating it is active.

In the file menu, click the **Multispectral Tab**.

Select "**Custom**" from the available Sensor list drop-down list within the Bands category grouping.

Experiment with various band combinations by assigning different bands to each of the colors (red, green, and blue) to achieve a FCC image (for example, vegetation is red, and water is dark blue/black) that resembles the Sentinel-2 FCC display in the second 2D View #2 graphic viewer window. Use the **Zoom In** tool 🔍 as necessary (Figure 2.10).

Create a TCC display band combination:

Now repeat the previous steps to create a TCC band combination for the Sentinel-2 image (**T23KPQ_A008672_20170218T130353_b2-4_8**) displayed

Introduction to Image Data Processing 65

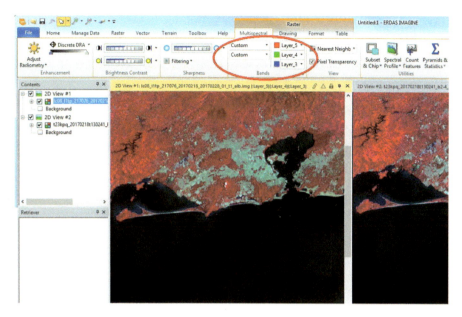

FIGURE 2.10
Landsat 8 FCC band combination selection (left 2D View display). (From ERDAS IMAGINE®/Hexagon Geospatial.)

in the second 2D View window. However, to create the TCC band combination it is necessary to use a 3 (red), 2 (green), 1 (blue) band combination from the multi-band image. Remember to click on the second 2D View window, turning the frame yellow, to activate it as pictured in Figure 2.11.

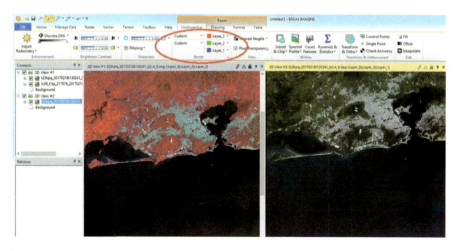

FIGURE 2.11
Sentinel-2 TCC band combination selection (right 2D View display). (From ERDAS IMAGINE®/Hexagon Geospatial.)

66 *Image Processing and Data Analysis with ERDAS IMAGINE®*

Linking Images in Multiple 2D Viewer Windows

Two or more viewers may be linked together to create a linked image. With the linked image active, identical locations on each images within each viewer will update simultaneously.

Link Multiple Viewers:

Make the **2D View #1** graphic window active

From the **Home** tab, choose the Link Views ⬚ option.

Now zoom and pan around to different locations on the image in **2D View #1** and notice the result on the image in **2D View #2**.

Click on the **Inquire Cursor** button ⬚ Inquire. White crosshairs will appear in both images, confirming that the images are now linked. The pixel location at the intersection of the crosshairs will be displayed in a viewer window.

Using the **Select Tool** ⬚ Select the crosshairs may be moved simultaneously in both images with your mouse.

Opening Multiple Images

Multiple images can be opened in a single viewer window.

Open Multiple Images in a single viewer:

Close the **Inquire Cursor View #1** window and deactivate both the **Inquire** and **Link Views** options.

Activate the **2D View #1** window.

From the **Open Folder** icon ⬚ at the top of the ERDAS window, make the follow selections:

- In the **Select Layer to Add** dialog box, highlight the Sentinel-2 **(T23KPQ_A008672_20170218T130353_B2-4_8.img)** image
- click on the **Raster Options** tab
- In the Layers to Color section enter the create the band combination 3 (red), 2 (green), 1 (blue) to create a TCC band combination
- Make sure the "**Clear Display**" box is unchecked
- Click on "**OK**" (Figure 2.12)

NOTE: Alternatively, you could drag the **T23KPQ_A008672_20170218T130353_ B2-4_8.img** from the Contents area into the **2D View #1** window.

The Sentinel-2 **(T23KPQ_A008672_20170218T130353_B2-4_8.img)** image is now added to the **2D View #1** window, on top of the Landsat 8 **(LC08_L1T P_011062_20160503_20170325_01_T1_ALB)** image.

NOTE: In the Contents area (under 2D View #1) it is also possible to change the order of the way the images are displayed in 2D View #1 so that the

Introduction to Image Data Processing

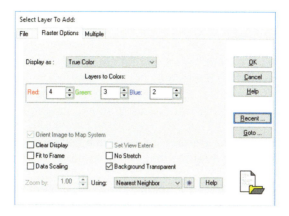

FIGURE 2.12
Select Layer to Add dialog window for creating a TCC band combination for the Sentinel-2 dataset. (From ERDAS IMAGINE®/Hexagon Geospatial.)

Landsat 8 **(LC08_L1TP_011062_20160503_20170325_01_T1_ALB)** image is on top of the Sentinel-2 **(T23KPQ_A008672_20170218T130353_B2-4_8.img)** image.

Select the *Swipe* button on the **Home** menu tab (**Home** tab | **View** grouping | **Swipe**), and then move the *Transition Extent* slider back and forth. This allows a quick and easy comparison between the top Sentinel-2 image and underlying Landsat 8 image (Figure 2.13).

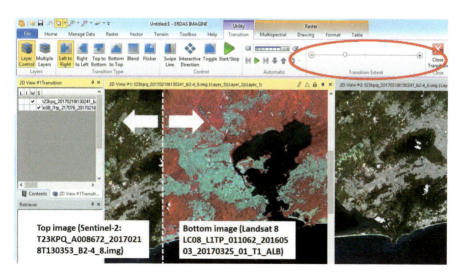

FIGURE 2.13
ERDAS Image layer transition swipe tool. (From ERDAS IMAGINE®/Hexagon Geospatial.)

Review Questions

1. True or false: A number of digital image analysis packages are currently available. Some will be more feature-packed and user-adaptable than others. However, any investigator should analyze their current available options as it relates to their own specific needs.

2. True or false: Freeware and open source packages are becoming more available and highly competitive to many of the commercial packages.

3. True or false: USGS EarthExplorer (http://earthexplorer.usgs.gov/), operated by the US government maintains a highly comprehensive archive of raster-based, remotely sensed data from all over the world.

4. When searching for imagery within the USGS EarthExplorer web portal, how can the Show Metadata and Browse icon ▣ be used to determine whether the recorded image quality meets the specified criteria necessary?

5. What types of information may typically be available from reviewing the image metadata ▣ in most digital image processing software packages?

6. In ERDAS IMAGINE, as well as other software image processing packages, the image displayed is determined by the combinations of satellite image-spectral bands used for the display and the three additive primary colors (red, green and blue) that are associated with those bands. Each image spectral band has captured a different part of the electromagnetic spectrum. What would it mean to display a "True Color" or natural color image? What would it mean to display a "False Color" image?

7. When creating a multi-band image using a layer stack operation, why should only the single bands that possess similar spatial resolutions be used?

8. The "**Swipe**" button ▣ Swipe, introduced in the "Opening Multiple Images in a Single Window" section of Chapter 2, provides some very useful functionality. In what ways this tool can be used when comparing imagery?

3

Georectification

Overview

Geometric distortions are errors common to any type of remotely sensed data, as various degrees of distortions are characteristic of the attempt to portray a three-dimensional surface onto a two-dimensional image. Geometric image distortions are generally divided into two types of errors, systematic and nonsystematic errors. Systematic distortions are due to image motion caused by the forward movement of the spacecraft, variations in the mirror scanning rate, panoramic distortions, variations in platform velocity, and distortions due to the rotation as well as the curvature of the Earth. These types of errors may be compensated for, reduced, or corrected, on board the aircraft or satellite, or through preprocessing using standard equations. Because these types of errors are assumed to be systematic (roughly affecting the entire image equally), applied correction procedures will systematically function to adjust all pixel locations throughout the image. Nonsystematic distortions are typically due to variations in satellite altitude, speed, and angular orientation in reference to the ground (attitude) and may be more difficult to identify and remove or reduce. The most common techniques for removing the remaining systematic and nonsystematic distortions are image-to-map rectification and image-to-image registration through the selection of a large number of ground control points.

In addition to geometric distortions, often it is important to georeference digital images. This process simply relates the images to the Earth's surface and assign geographic coordinates. As well this procedure is useful for minimizing alignment errors that could occur when using multiple datasets. There are several different ways to georeference digital images. Two common methods are rubber sheeting and the use of polynomial coordinate transformation equations. These methods are referred to as georeferencing or georectification. The georectification process uses equations to assign or rectify map, or X-Y location coordinates to the image. Polynomial transformation (linear 1st order or affine transformation) may be used to adjust location, scale, and skew error in the X or Y directions, as well as image rotation. These transformations are usually based on rectified map coordinates

69

captured in the field—collected using GPS-control points—or directly from an image of the same area that has been previously rectified to a particular map projection of interest. Rubber Sheeting transformations are used to correct nonlinear distortions, such as distortions in data covering large areas, to account for the curvature of the Earth.

Image Preprocessing—Georectification

In this exercise, adapted from original research developed by Dr. Heather Cheshire at North Carolina State University, we will rectify an unprojected image of the Carl Alwin Schenck Memorial Forest (Schenck.img) using an image of the Schenck Memorial Forest that has a known projection (Schenck_doq.img) as a reference. This kind of rectification is known as image-to-image rectification. The Schenck_doq.img image was rectified using the State Plane map projection. For more information on the Carl Alwin Schenck Memorial Forest, located in Raleigh, North Carolina, visit the following link: https://www.ncsu.edu/scilink/studysite/data_files/schenckforest.html.

When rectifying the Schenck.img image follow these six basic steps: (1) display images, (2) start Geometric Correction Tools, (3) record Ground Control Points (GCPs), (4) compute a transformation matrix, (5) resample the image, and (6) verify the rectification process. Two methods are used to complete these steps: Polynomial Regression and Rubber Sheeting.

Learning Objectives

1. Become familiar with the ERDAS IMAGINE® software interface through an introductory image processing technique.

2. Learn to georectify images using multiple techniques of image to image rectification.

3. Explore differences in Polynomial Regression and Rubber Sheeting methods.

Data required: For this exercise, the following image files will be used and should be located on the companion media, or within your default working directory (see Chapter 2: Setting up workspace preferences):

1. **Schenck.img**—An uncorrected digital image of the Schenck area in Imagine format (*.img)

2. **Schenck_doq.img**—A Digital Orthophoto Quarter Quadrangle (DOQQ) from the same area that will be used to provide horizontal reference (x- and y-coordinates) for the GCPs

Georectification 71

Rectifying Image of Schenk Forest Using Polynomial Regression and Rubber Sheeting

I. Polynomial Regression

Display Images—Start Two Viewers (File | New | 2D View)

1. In the left viewer open the **Schenck.img** image to be rectified (**File | Open | Raster Layer**). This will serve as the input image.
2. In the right viewer open the **Schenck_doq.img** image. This will serve as the reference image.
3. Fit images to frames by selecting each 2D Viewer and right-clicking "**Fit to Frame**" icon in the file menu ribbon as seen in the following image (Figure 3.1).

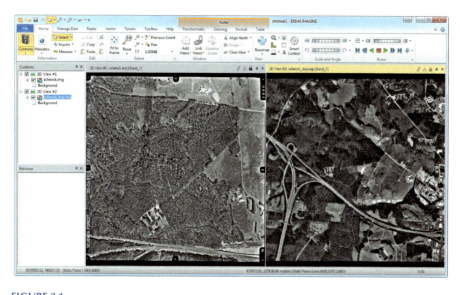

FIGURE 3.1
ERDAS IMAGINE image 2D Viewer #1 and #2 display. (From ERDAS IMAGINE®/Hexagon Geospatial.)

Start Geometric Correction Tools

1. For the Viewer #1, choose **Panchromatic | Transform & Orthocorrect tools | Control Points**.
2. Under Select Geometric Model, set geometric model to **Polynomial** and click **OK** (Figure 3.2).

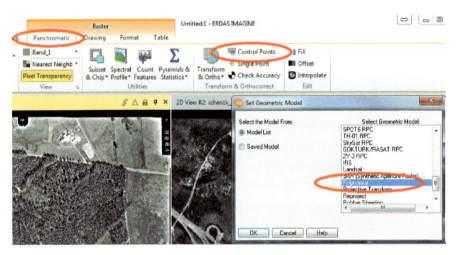

FIGURE 3.2
ERDAS IMAGINE select geometric model. (From ERDAS IMAGINE®/Hexagon Geospatial.)

3. The **GCP Tool Reference Setup** window opens. The **GCP Tool Reference Setup** dialog specifies the various ways that reference GCPs can be collected. Reference GCPs are reference points for which real-world coordinates are known. Reference GCPs will be obtained from the **schenck_doq.img** image, so select the **Image Layer** radio button and **OK** (Figure 3.3).

Georectification 73

FIGURE 3.3
GCP Tool reference setup. (From ERDAS IMAGINE®/Hexagon Geospatial.)

4. Navigate to the reference image **schenck_doq.img** and select it. The **Reference Map Information** window should appear with the current coordinate reference system (for the reference image). Review this information and click **OK** when done (Figure 3.4).

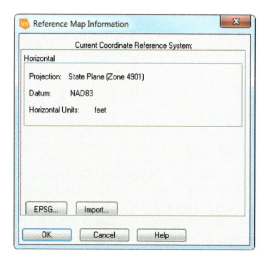

FIGURE 3.4
Reference map information. (From ERDAS IMAGINE®/Hexagon Geospatial.)

5. Notice that in the **Polynomial Model Properties** dialog, the *Status* line at the bottom reads "**Model has no solution**"—when adding GCPs and calculating a transformation matrix, it will change. For now, do not close the Polynomial Model Properties dialog, simply minimize this dialog (Figure 3.5).

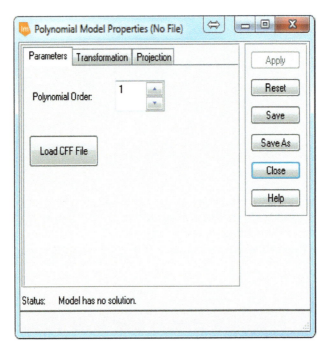

FIGURE 3.5
Polynomial model properties. (From ERDAS IMAGINE®/Hexagon Geospatial.)

Record Ground Control Points

You should now see that the **Multipoint Geometric Correction** dialog opened. The left side of the viewer shows **schenck.img** in its full extent (top left), a GCP 'zoom' viewer (top right) and a chip extraction viewer (bottom). The right side of the viewer shows the **schenck_doq.img** in its full extent, a GCP 'zoom' viewer and a chip extraction viewer (Figure 3.6).

Georectification

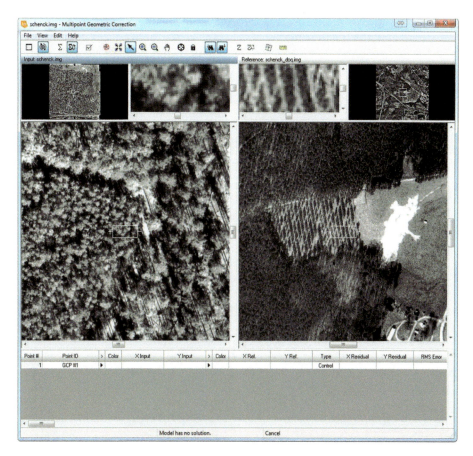

FIGURE 3.6
Multipoint geometric correction dialog. (From ERDAS IMAGINE®/Hexagon Geospatial.)

1. Try moving or resizing a link box in each viewer (make sure the pointer tool is selected in the GCP Tool window, otherwise you'll add a GCP) and the corresponding chip extraction viewer will pan and rescale. The purpose of these two additional windows is to zoom in on some part of the image in order to position a GCP precisely.

2. Next add four pairs of GCPs (only three for polynomial transformation of the first order, but four or more is better). A pair consists of an input GCP (recorded in the X Input and Y Input columns of GCP Tool) and a corresponding reference GCP (recorded in the X Ref., Y Ref. columns). To create GCPs, look for locations that are (a) present in both images, (b) easily and accurately identifiable in both images, and (c) evenly spread inside images, preferably closer to image edges.

3. As a starting point, use the cleared land area just above the three roads of the visible highway system in the bottom-left quadrant of schenck.img, and mid-left of schenck_doq.img as seen in the following image. Use the top-right corner of this area for GCP #1. Drag both link boxes to it, then resize the boxes and/or chip extraction viewers so that you can clearly see the detail of the corner (Figure 3.7).

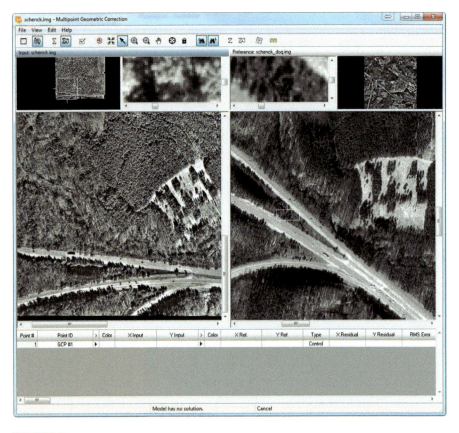

FIGURE 3.7
Multipoint geometric correction control point location. (From ERDAS IMAGINE®/Hexagon Geospatial.)

4. Now switch to **Create GCP** tool in the **GCP Tool bar**. Add a GCP in the input image (Schenck.img), then in the reference image (Schenck_doq.img).

NOTE: You need to select the GCP tool again to add a GCP in the reference image.

Georectification

You can select and change the Color option in the table to make GCPs easier to distinguish from one another (e.g., GCP # 1 was changed to **red color**) (Figure 3.8).

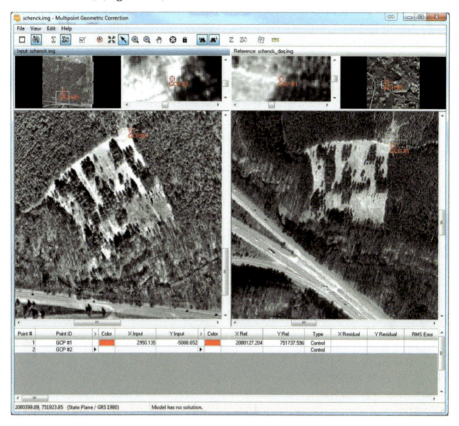

FIGURE 3.8
Multipoint geometric correction control point color. (From ERDAS IMAGINE®/Hexagon Geospatial.)

5. Add at least three or more pairs of GCPs to different locations throughout the image.

   NOTE: After adding a fourth GCP to the input image a corresponding GCP will automatically be created in the reference image.

   GCPs should be spaced as widely as possible throughout the image, with the best placement placement of the points as close to edges or corners as possible. The wide distribution of GCPs ensures that the calculation of polynomial transformation model will be valid for the entire image and not just the areas where GCPs are concentrated. Notice that the **Schenck_doq.img** is smaller in scale than **Schenck.img** so spread out GCPs as much as possible in **Schenck.img**.

NOTE: Use caution when adding GCPs so that corresponding points aren't added to another row incorrectly—the pointers ">" show the current row for input and reference GCPs. A GCP can be moved around the image by dragging it to a new location. To delete an unwanted GCP, simply right-click on the corresponding row within the Point # column and select Delete Selection from the pop-up dialog.

After at least four pairs of GCPs have been collected, review the Control Point Error field in the button bar of the GCP Tool. The error is given in input units, which in this case is in pixels. The **Total error** should probably be less than 10 units (this is a good starting point, however as little error as possible is always better). Next, practice reducing your total error to as low as possible.

NOTE: Total error of <0.5 is considered good in most professional industries (Figure 3.9).

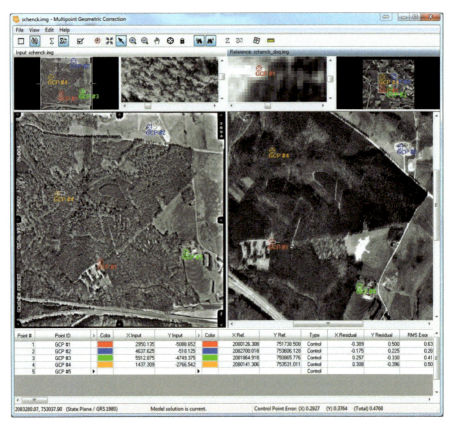

FIGURE 3.9
Multipoint geometric correction total error. (From ERDAS IMAGINE®/Hexagon Geospatial.)

Georectification 79

6. Save the GCPs you created for use in the next exercise. You have to save each GCP individually, so you will be saving two different GCP files (*.gcc).

   To save the input image GCPs simply select "**File | Save Input as…**" and name the file **Schenck_input**. Now save the reference GCPs by selecting "**File | Save Reference As…**" and name the file **Schenck_ref**. You will use each of these in the next exercise so remember where you saved them.

7. The table generated from the GCP Tool can be saved as a **Report**. First clear all previously selected rows in the table (right-click under Point # column on *1* and then choosing Select None). Once all selections have been cleared, right-click on column header Point ID and drag cursor to select all. With all selected, right-click on **Point ID** column heading and choose **Report…** In the **Report Format Definition** dialog, select **OK** (Figure 3.10).

FIGURE 3.10
Report format definition. (From ERDAS IMAGINE®/Hexagon Geospatial.)

8. Next, the **Editor** dialog will open and display a copy of the **Error Report** (Figure 3.11).

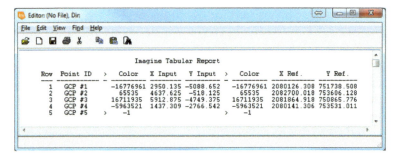

FIGURE 3.11
Error report. (From ERDAS IMAGINE®/Hexagon Geospatial.)

9. Save the report as "Schenck GCPs Error Report" to your home directory (This is a text file, you can open it with Word to see it in the proper format). To do so, select **File | Save As...** to save a copy of this Error Report. Once you have navigated to your working directory, choose the file name **Schenck GCPs Error Report** and next select **ASCII Text File (*.txt)** as the type of file format to save the report in. Next click **OK**.

Compute Transformation Matrix

In the **Multipoint Geometric Correction** dialog, click on **Display Model Properties** (the first icon ▣) to verify that the Status is now "Model solution is current" in the **Polynomial Model Properties** dialog. Close the window (Figure 3.12).

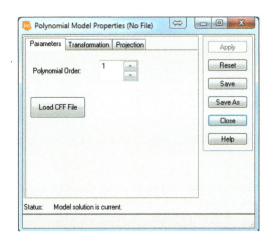

FIGURE 3.12
Polynomial model properties showing the updated status that the model solution is now current. (From ERDAS IMAGINE®/Hexagon Geospatial.)

Resample the Image

1. In the **Multipoint Geometric Correction** dialog, click on the **Resample Image Dialog** button (next to last beside ruler). Name the Output File as **schenck_poly.img** and save this file to your home directory (i.e., click on the little yellow directory folder). Leave all other as default. Use the **Nearest Neighbor** sampling technique to maintain the integrity of the original image (other methods may distort original values) (Figure 3.13).

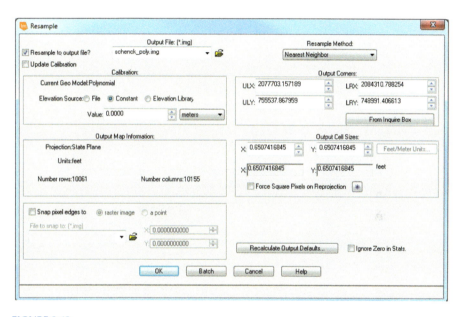

FIGURE 3.13
Resample image dialog. (From ERDAS IMAGINE®/Hexagon Geospatial.)

2. Click **OK** to start resampling. The **Process List** should appear. Close after processing is finished.
3. Save the geometric model by opening the **Polynomial Model Properties** box you minimized earlier (or clicking the **Display Model Properties** icon and select **Save As**. Save the model as **Schenck_model.gms**.

NOTE: You will not be able to save by simply clicking "**File | Save As...**" in the Multipoint Geometric Correction dialog box. You **MUST** save from the **Polynomial Model Properties** box. If you do not save, the model **WILL NOT** be saved automatically (Figures 3.14 and 3.15).

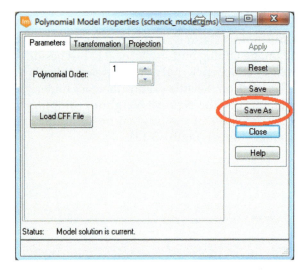

FIGURE 3.14
Polynomial model properties. (From ERDAS IMAGINE®/Hexagon Geospatial.)

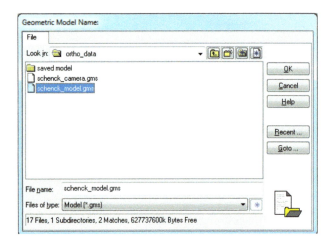

FIGURE 3.15
Geometric model name. (From ERDAS IMAGINE®/Hexagon Geospatial.)

Verify Rectification

1. Close the **Multipoint Geometric Correction** interface and return to viewers #1 and #2.

2. In Viewer #2 (currently displaying **schenck_doq**), open the **schenck_poly.img** image (you may close viewer #1).

Georectification 83

3. Right-click on the layer and select **pixel transparency** (to turn background pixels off) and **Fit Layer to Window** (notice the image is now tilted, which signifies it now conforms to a map projection) (Figure 3.16).

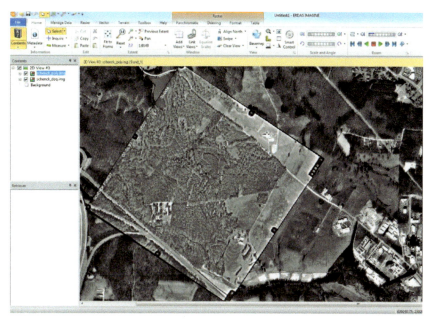

FIGURE 3.16
ERDAS IMAGINE rectification display. (From ERDAS IMAGINE®/Hexagon Geospatial.)

4. Visually evaluate the result of rectification by using the Viewer Swipe tool (**Home/Swipe**). Drag the Swipe Position lever back and forth. The images should align (Figure 3.17).

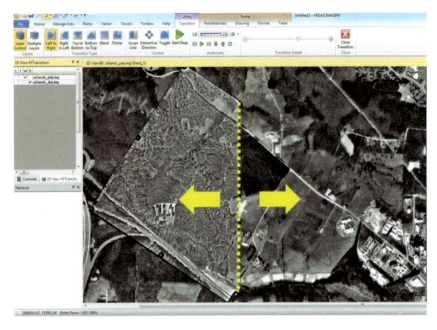

FIGURE 3.17
ERDAS rectification swipe comparison of model results. (From ERDAS IMAGINE®/Hexagon Geospatial.)

Georectification 85

5. Now clear the ERDAS IMAGINE graphic view window by right-clicking in the view window and selecting "**Clear View**" (Figure 3.18).

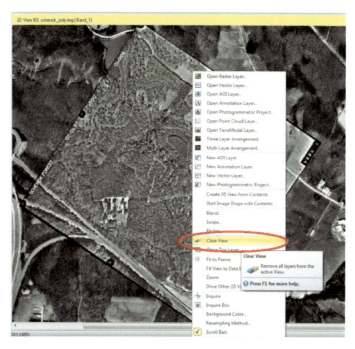

FIGURE 3.18
Clear view. (From ERDAS IMAGINE®/Hexagon Geospatial.)

II. Rubber Sheeting

Display Images

1. Start two viewers (**File | New | 2D View**).
2. In the left viewer, open the **Schenck.img** image to be rectified (**File | Open | Raster Layer**). This will serve as the **input image**. (A common trick is to choose **File | Recent** and find the original files if you have worked with them before).
3. In the right viewer, open the **Schenck_doq.img** image. This will serve as the **reference image**.
4. Fit images to frames by selecting each 2D Viewer and right-clicking "**Fit to Frame.**" (Figure 3.19).

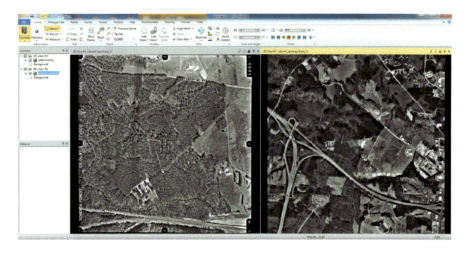

FIGURE 3.19
ERDAS IMAGINE rubber sheeting display setup. (From ERDAS IMAGINE®/Hexagon Geospatial.)

Georectification 87

Start Geometric Correction Tools

1. For Viewer #1, choose **Panchromatic | Transform & Orthocorrect tools | Control Points**.
2. Under **Select Geometric Model**, set geometric model to **Rubber Sheeting** and click **OK** (Figure 3.20).

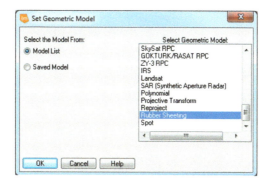

FIGURE 3.20
Set geometric model-Rubber sheeting. (From ERDAS IMAGINE®/Hexagon Geospatial.)

3. The **GCP Tool Reference Setup** window opens.

   The **GCP Tool Reference Setup** dialog lists all the ways in which reference GCPs could be obtained in Imagine. The **schenck_doq.img** image will be used, so select the **Image Layer** radio button and click **OK**.

4. Navigate to the reference image **schenck_doq.img** and select it. The **Reference Map Information** window should appear with the current coordinate reference system (for the reference image). Click **OK**.

5. For now, do not close the **Rubber Sheeting Model Properties** dialog box which opens after closing the **Reference Map Information** window.

Ground Control Points

1. Use the GCPs you created in the previous polynomial regression exercise (earlier). Load the reference GCPs you saved earlier by selecting **File | Load Reference GCPs** in the **Multipoint Geometric Correction** dialog box. Select the **Schenck_ref.gcc** file you saved earlier.

2. You will be prompted to define the reference map coordinate system. Select the **Current** system which will keep the current coordinate system. Click **OK** (Figure 3.21).

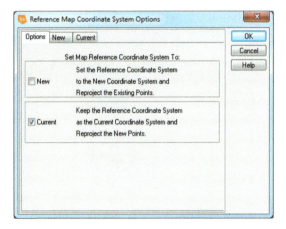

FIGURE 3.21
Reference map coordinate system options. (From ERDAS IMAGINE®/Hexagon Geospatial.)

3. Next, load the input GCPs. To do this, select **File | Load Input GCPs** in the **Multipoint Geometric Correction** dialog box. Select the **Schenck_input.gcc** file you saved earlier.
4. Now the input GCPs and reference GCPs should be loaded. You should not have to create ANY additional ground control points (Figure 3.22).

Georectification 89

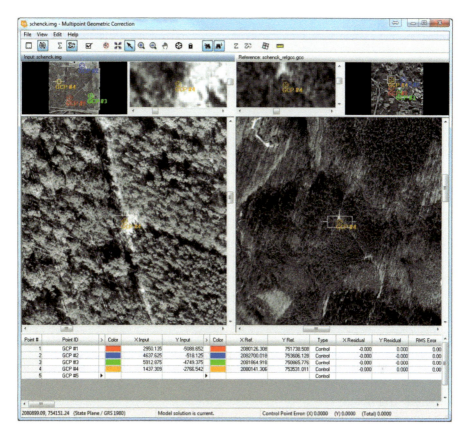

FIGURE 3.22
Input GCPs and reference GCPs. (From ERDAS IMAGINE®/Hexagon Geospatial.)

Compute Transformation Matrix

1. In the **Multipoint Geometric Correction dialog**, click on **Display Model Properties** (the first icon).
2. Verify that the Status is now "Model solution is current" in the **Model Properties** dialog.
3. Close the **Model Properties** dialog (Figure 3.23).

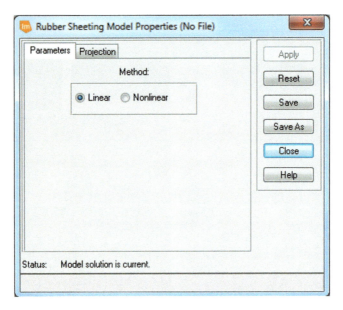

FIGURE 3.23
Rubber sheeting model properties dialog. (From ERDAS IMAGINE®/Hexagon Geospatial.)

Resample the Image

1. In the **Multipoint Geometric Correction** dialog, click on the **Resample Image Dialog** button (next to the last tool on the menu, beside ruler).

2. Name the Output File as **schenck_rubb.img** and save this file to your home directory (click on the little yellow directory folder).

   Use the **Nearest Neighbor** sampling technique to maintain the integrity of the original image (other methods may distort original values). Leave all other options as the default. Click **OK** to start resampling. The **Process List** should appear. **Close** the Process List after processing is finished.

3. Save the geometric model by opening the **Rubber Sheeting Model Properties** box, which you minimized earlier (alternatively click on **Display Model Properties** icon) and select **Save As** to save the model as **Schenck_model2.gms**.

Georectification   91

> NOTE: You will not be able to save by simply clicking "**File | Save As...**" in the **Multipoint Geometric Correction** dialog box. You **MUST** save from the **Rubber Sheeting Model Properties** box. If you do not save, the model **WILL NOT** be saved automatically.

Verify Rectification

1. Close the **Multipoint Geometric Correction** interface and return to viewers #1 and #2.

2. In Viewer #2 (currently displaying **schenck_dog.img**), open the **schenck_rubb.img** image on top of the current image in the viewer (schenck_doq.img) by right-clicking in **2DViewer #2| Open Raster...** | and navigating to where you saved the **schenck_rubb.img** image (you may close viewer #1).

3. Right-click on the layer in the Contents Legend and select **pixel transparency** (to turn off background pixels if not already off), and **Fit Layer to Window**. Notice the image is now *stretched*, which signifies it now conforms to a map projection (Figure 3.24).

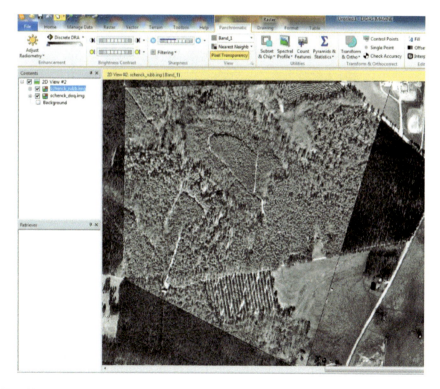

FIGURE 3.24
Completed rubber sheeting model results for the Schenck image. (From ERDAS IMAGINE®/Hexagon Geospatial.)

4. Visually evaluate the result of rectification by using the **Viewer Swipe tool** (**Home/Swipe**). Drag the Swipe Position lever back and forth. The images should align (Figure 3.25).

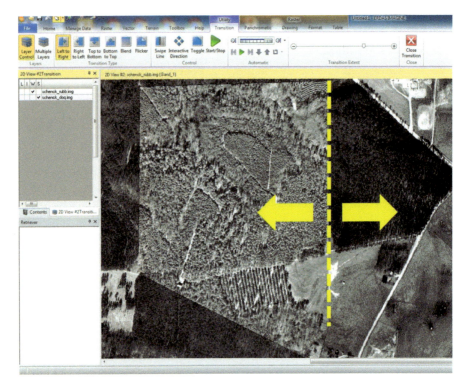

FIGURE 3.25
Swipe comparison of rubber sheeting model results. (From ERDAS IMAGINE®/Hexagon Geospatial.)

5. Close Swipe tool.
   Compare Geometric Correction Model Results: Polynomial vs. Rubber Sheeting
6. Add a new 2D View by selecting **file | New | 2-D View**.
7. Add **schenck_doq.img** to the new graphics viewer.
8. In the new viewer (currently displaying **schenck_dog.img**), open the **schenck_poly.img** image on top of the current image in the viewer (**schenck_doq.img**) by **right-clicking in 2DViewer #2 | Open Raster…** | and navigating to where you saved the **schenck_poly.img**.

Georectification

9. Right-click on the layer in the Contents Legend, select **pixel transparency** (to turn off background pixels if not already off), and **Fit Layer to Window** (Figure 3.26).

FIGURE 3.26
ERDAS IMAGINE display of rubber sheeting model result (a) and polynomial model result (b). (From ERDAS IMAGINE®/Hexagon Geospatial.)

Review Questions

1. Geometric image distortions are generally divided into two types of errors, systematic and nonsystematic errors. Explain these two types of errors.
2. The georectification processing application introduced two georectification models, the polynomial model and the rubber sheeting model. Each operation provides some advantages and disadvantages regarding final output. For the Polynomial model what are possible advantages and disadvantages that may be noticeable in the final georectified image?
3. For the rubber sheeting model what are possible advantages and disadvantages that may be noticeable in the final georectified image?
4. In what situation might the polynomial model be beneficial over the rubber sheeting model?
5. In what situation might the rubber sheeting model might be beneficial over the polynomial model?

4

Orthorectification

Overview

In the previous exercise, an uncorrected digital image was georeferenced to a reference image through two common methods; **Rubber Sheeting** and the use of **Polynomial Coordinate Transformation** equations. **Orthorectification** is different from these two procedures in that it uses z-coordinate values (elevation data) in the georeferencing process. The use of elevation data in orthorectification allows for the correction of relief displacement. Orthorectification also incorporates camera-specific information to correct certain optical displacement effects.

The output of orthorectification is an orthophoto. Orthophotos are planimetrically accurate, meaning scale is constant across the image. This allows orthophotos to be used as base maps, which they commonly are in a wide variety of Geographic Information Systems (GIS) applications.

Image Preprocessing—Orthorectification

In this exercise, adapted from original research developed by Dr. Heather Cheshire at North Carolina State University, you will use the ERDAS IMAGINE® software to orthorectify a single digital image of the Schenck Memorial Forest. You will define interior orientation based on the relative positions of the fiducial marks as well as information from the camera calibration report. Then, you will develop a geographic *solution* (relate image coordinates to ground coordinates) based on a system of ground control points (GCPs). Once you have minimized the root mean square error (RMSE) associated with the GCPs, you will generate a new orthophoto image based on this solution.

96   *Image Processing and Data Analysis with ERDAS IMAGINE®*

This single-image approach is fairly simple compared to aerial triangulation procedures for processing an entire block of photos all at once. For example, radial lens distortion will not be an issue, nor will tie points be required in a single-image approach. Still, this exercise will provide experience with some of the fundamental techniques of digital photogrammetry.

Learning Objectives

1. Increase your familiarity with ERDAS IMAGINE image processing.
2. Learn to orthorectify a digital image of the Schenck Memorial Forest.
3. Explore fundamental techniques of digital photogrammetry.

Data required: You will need the following data for this exercise:

1. **Schenck camera calibration report**—camera #142821, lens #142829 (top-right corner of the first page)
2. **Schenck.img**—An uncorrected digital image of the Schenck area in Imagine format
3. **Schenck_doq.img**—A Digital Orthophoto Quarter Quadrangle (DOQQ) from the area that will be used to provide horizontal reference (i.e., x- and y-coordinates) for the GCPs
4. **Schenck_lidar20ft.img**—A Digital Elevation Model (DEM) from the area that will provide vertical reference (z-coordinates) for the GCPs

---

NOTE: The projection of the reference data, and thus the projection for this exercise, is as follows:

Projection: **U.S. State Plane**
Zone: **4901 (FIPS zone 3200)—North Carolina**
Datum: **NAD 83**
Units: **Feet**
Spheroid: **GRS 80**

Getting Started—Orthorectification

1. Launch the ERDAS IMAGINE software. The Imagine window will appear along with **2D View #1.**
2. Right-click on in the View #1 window and select **File | Open | Raster Layer** from the menu. Navigate to the directory with the image **schenck.img**. Select the image and click **OK** to open it in the viewer.

Orthorectification

>   NOTE: To review the two reference files, **schenck_doq.img** and **schenck_lidar20ft.img**, launch a second 2D view (#2) and open the images in that window (click on File in the upper left corner and select **New | 2D View**).

3. On the main Imagine window, click on **Panchromatic** (Make sure 2-D View #1 is highlighted in the viewer). Within the **Transform & Orthocorrect** grouping, select **Control Points**.
4. This will launch the **Set Geometric Model** dialog (Figure 4.1):

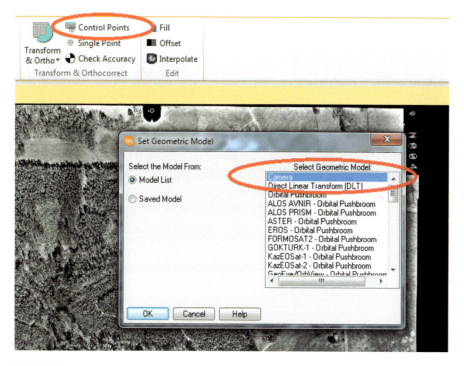

FIGURE 4.1
Set geometric model dialog. (From ERDAS IMAGINE®/Hexagon Geospatial.)

Notice that this dialog provides several options (some not shown in this figure), including the polynomial and rubber sheeting methods discussed previously in Chapter 3.

5. Select the **Camera** option and click **OK**.
6. This will open the Multipoint Geometric Correction window and GCP Tool Reference Setup window (Figure 4.2).

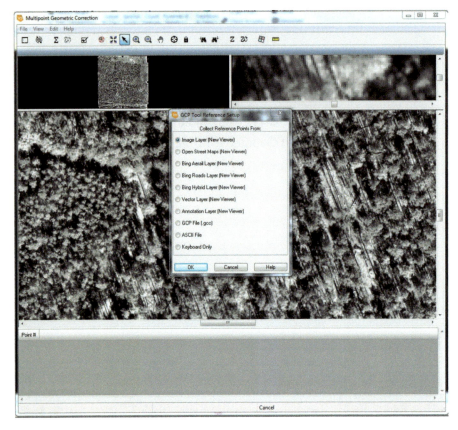

FIGURE 4.2
Multipoint geometric correction window and GCP tool reference setup window. (From ERDAS IMAGINE®/Hexagon Geospatial.)

7. In the GCP Tool Reference Setup dialog window and under the "Collect Reference Points From" options, ensure "**Image Layer** (New Viewer)" is selected, click "**OK**" to close the GCP Tool Reference Setup dialog window, and specify the **schenck_doq.img** image file in the Reference Image Layer dialog window the opens next and select "**OK**."

8. The "**Reference Map Information**" dialog window opens next and shows that the reference image projection (Horizontal) is State Plane, NAD83, feet. Click on "**OK**" (Figure 4.3).

Orthorectification

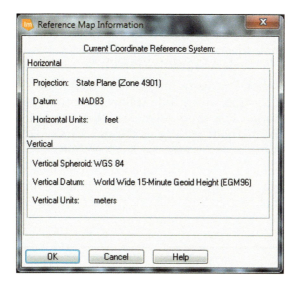

FIGURE 4.3
Reference map information dialog window. (From ERDAS IMAGINE®/Hexagon Geospatial.)

Defining Camera Properties (Interior Orientation)

1. After selecting **OK** to close the "**Reference Map Information**" dialog window, the **Camera Model Properties** dialog box opens (Figure 4.4).

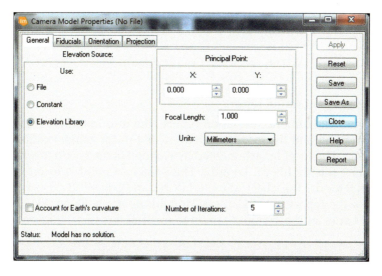

FIGURE 4.4
Camera model properties dialog. (From ERDAS IMAGINE®/Hexagon Geospatial.)

Notice that the Camera Model Properties dialog box has four "page" tabs: **General**, **Fiducials**, **Orientation**, and **Projection**. On the first page (General), there are several items displayed, some of which will come from the camera calibration report.

2. Under **Elevation Source** grouping, click on **File** and select the elevation file by clicking on the little yellow folder navigation icon next to the Elevation File drop-down list. Navigate to and select the file **schenck_lidar20ft_v2.img** in the download folder containing the data available for this exercise. Click on "**OK.**"

Should you receive a "**Warning on DEM file**" message, disregard this message. The supplied Light Detection and Ranging (LiDAR) data does not contain any NODATA values. Continue by selecting "**Yes**" (Figure 4.5).

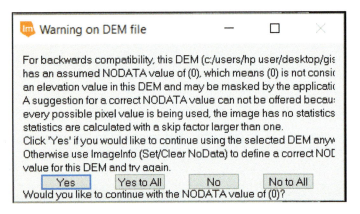

FIGURE 4.5
NODATA warning on DEM file. (From ERDAS IMAGINE®/Hexagon Geospatial.)

3. **Camera calibration report**

Review the camera calibration report. The information on the principal point and the focal length are listed in the camera calibration report. The **focal length** is listed in Section I of the report (next to the label Calibrated Focal Length). The **principal point** is listed in section II (under the label Calibrated Principal Point). Enter these values from the report to the Camera Model Properties dialog. Make sure the **Units** for these values are set to **millimeters** (mm) (Figure 4.6).

Orthorectification 101

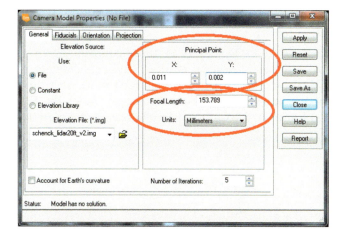

FIGURE 4.6
Camera model properties. (From ERDAS IMAGINE®/Hexagon Geospatial.)

The default settings are fine for the "Number of Iterations," "Account for Earth's curvature," and any of the other buttons or checkboxes. For further information on these settings, review the ERDAS IMAGINE Online Help.

4. Click on the **Fiducials** tab. The Schenck image has eight fiducial marks, so click on the **Fiducial Type** button marked with the numbers 1–8. This indicates you will be marking eight fiducial marks; it does not indicate the order of the fiducials (Figure 4.7).

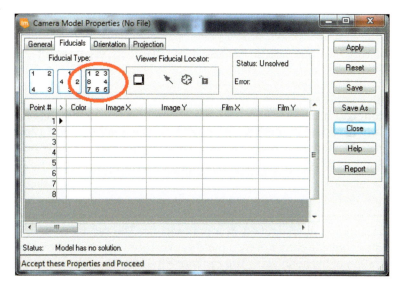

FIGURE 4.7
Camera model properties dialog. (From ERDAS IMAGINE®/Hexagon Geospatial.)

5. Next set the fiducial locations using information from the **camera calibration report**.

In Section VII, there is a list of x- and y-coordinates (in mm) associated with the eight fiducial marks. (These coordinates record the X and Y distances between each fiducial mark and the principal point of the image). List these coordinates under the **Film X** and **Film Y** columns in the Camera Model Properties dialog (Figure 4.8).

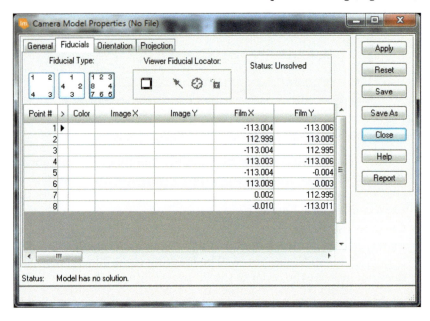

FIGURE 4.8
Camera model properties fiducial locations dialog. (From ERDAS IMAGINE®/Hexagon Geospatial.)

6. After you have entered these fiducial coordinates, save the **camera model** to a geometric model file (*.gms) before continuing in the **Camera Model Properties** dialog window (**save as | schenck_camera.gms**). Next, minimize the **Camera Model Properties** dialog window.

7. Switch back to the **Multipoint Geometric Correction** window by clicking on the heading of the **Multipoint Geometric Correction** dialog window (do **not** close the **Camera Model Properties** dialog!). Notice that the area enclosed by the target box (at the center of the crosshairs) in each main viewer is magnified in a mini-viewer. The second mini-viewer (upper outside corner of each main viewer) is an overview showing the location of the main view. You can move the target box elsewhere in the image by dragging the crosshairs with the cursor. You can also click on the target box itself and change its shape. This will allow you to zoom in on a smaller area or zoom out to a larger area (Figure 4.9).

Orthorectification

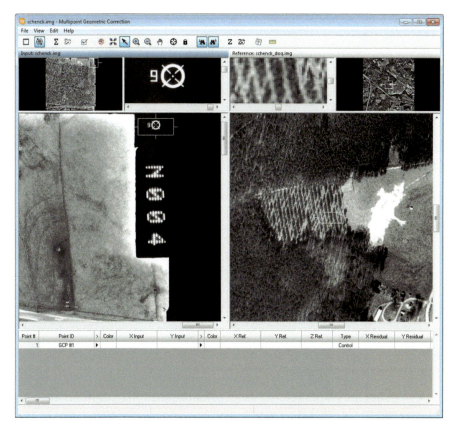

FIGURE 4.9
Multipoint geometric correction window. (From ERDAS IMAGINE®/Hexagon Geospatial.)

8. If you refer to Section VII of the camera calibration report, you'll notice that there is a *particular* numbering order for the fiducials, and that this relates to the "data strip side" of the image.

NOTE: Though the date, scale, and other data features are stamped on the left-hand side of the image, this is *NOT* really the data strip. Instead, the data strip usually lists specific information about the camera (serial number, possibly focal length, etc.), and is found along the black edge of a photo, not stamped in the image itself. Because there is a frame counter on the right-hand side of the image (found at the top-right corner), the right side will be called the data strip side.

This means that the fiducials **MUST** be marked on the image in the following order shown in Figure 4.10.

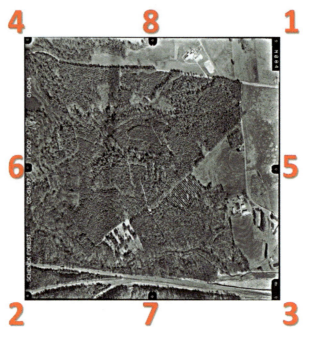

FIGURE 4.10
Fiducials numbering order to be marked on the image.

Using the camera report as an example, the fiducials order are as shown in Figure 4.11.

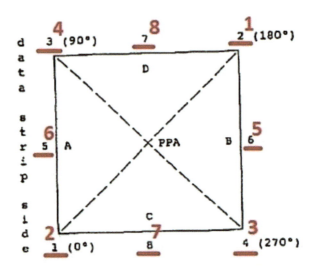

FIGURE 4.11
Camera report fiducials order (labeled in red).

Orthorectification

So, when you move the target box to the fiducial mark at the top right-hand corner of the image, you will be marking the fiducial labeled **1** in Section VII of the camera calibration report. When you mark the fiducial on the right-edge center, you will be marking the fiducial labeled **5** in Section VII of the camera calibration report, and so on. *Make sure you are labeling each point in the order shown in Figures 4.10 and 4.11 earlier!*

9. Move the target box to the upper-right corner of the image in the **schenck.img** (input image) in the **Multipoint Geometric Correction** window. Make sure **point #1** is selected (be sure to click on the appropriate row under ">" column; the corresponding row should have a pointer "▶" marking the designation row of the value to input) (Figure 4.12).

FIGURE 4.12
Pointer locations for adding ground control points (GCP). (From ERDAS IMAGINE®/Hexagon Geospatial.)

10. In the **Camera Model Properties** dialog, under **Viewer Fiducial Locator**, click on "Toggle viewer selectors for image fiducial input" icon to activate the three adjacent buttons. Click on the **Place Image Fiducial** button (Figure 4.13).

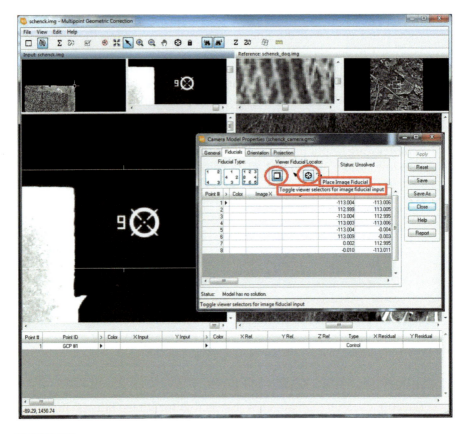

FIGURE 4.13
Toggle viewer selectors for image fiducial input and Place Image Fiducial buttons. (From ERDAS IMAGINE®/Hexagon Geospatial.)

11. Switch back to the **Multipoint Geometric Correction** window. In the upper-right corner of the close-up/zoom viewer window, place the cursor as close as possible to the exact center of the fiducial mark (Figure 4.14). (If you want to see the placed point marker better, you can change the color of the point marker in the Camera Model Properties dialog.)

Orthorectification

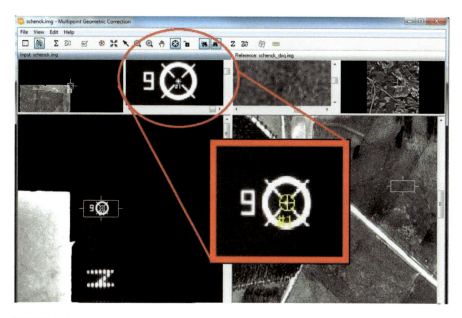

FIGURE 4.14
Zoomed in location of the exact center of the fiducial mark. (From ERDAS IMAGINE®/Hexagon Geospatial.)

12. Now move the magnifier box in the **Multipoint Geometric Correction** window for the **schenck.img** input image to the middle fiducial of the right side. Back in the **Camera Model Properties** dialog box, make sure **point #5** is selected (pointer "▶" is displayed in that row) and the **Place Image Fiducial** button is active before marking this point on the image. Repeat this process for the other six fiducial marks on the image. Remember to **do them in the order described** (refer to the numbers in red on the aforementioned Figure 4.11 *Camera report fiducials order*) (Figure 4.15).

108 *Image Processing and Data Analysis with ERDAS IMAGINE®*

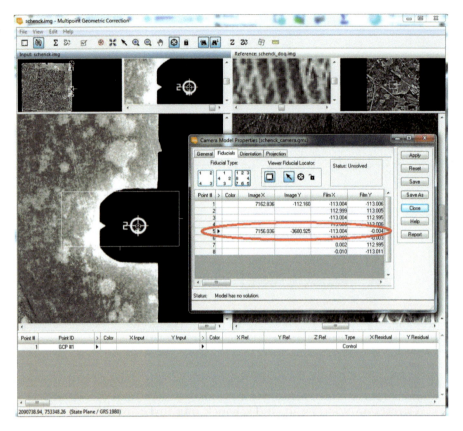

FIGURE 4.15
Pointer locations for adding Point #5. (From ERDAS IMAGINE®/Hexagon Geospatial.)

13. After all the fiducials have been placed, you will notice that an error value shows up in the **Camera Model Properties** dialog (Figure 4.16).

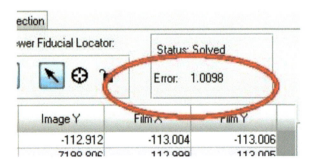

FIGURE 4.16
Camera model properties error dialog. (From ERDAS IMAGINE®/Hexagon Geospatial.)

Orthorectification 109

NOTE: Why is there an error? Remember, among other things (exact placement of the fiducial points etc.), there is also some distortion from the scanning process.

14. You can reduce this error by adjusting the position of the fiducial marks. If you click on and highlight one of the fiducial rows in the **Camera Model Properties** dialog, the target box will move to that fiducial mark, and the mark will be displayed in the close-up viewer. You can make the target box smaller and zoom in even closer in the close-up viewer. Then, you can move the point marker using the cursor.

    Chances are if you were careful when placing the original points, then the residual error will be very small (ideally, the smaller the better; typically, around 0.5000 is generally the rule of thumb). Still, the goal is to get the error less than 10.0000. Manipulate the fiducial marks until you get a smaller error value (very close to or below 10.0000).

15. There are two additional page tabs in the Camera Model Properties dialog (Orientation and Projection). Skip the Orientation tab (this is used when exterior orientation information is available from airborne GPS or inertial measurement unit (IMU)) and go to the Projection tab.

    Click on the **Projection** tab. Under the **Horizontal** area you should have—Projection: **State Plane (Zone 4901)**, and Datum: **NAD 83**. Make sure the Horizontal units are set to **feet**. Under the **Vertical**, by clicking the "**Set…**" button, you should have options to select the Spheroid name (**GRS 1980**), and Datum name **(NAD 83)** from the corresponding dropdown lists within the Elevation Info Chooser dialog window that opens.

16. After clicking **OK**, the updated projection information should be displayed in the Camera Model Properties dialog.

17. Finally, save the camera model (click the **Save** button is in the Camera Model Properties dialog, or Save As to ensure you have saved the model as **schenck_camera.gms**). Then **close** the Camera Model Properties dialog.

Selecting Ground Control Points

The **GCP Tool** in the **Multipoint Geometric Correction** window, in many ways, works similarly to the Fiducial Locator you used in the Camera Model Properties dialog in the previous section.

1. Again, you will use the target boxes on the two images and mini-viewers for close-ups. You'll notice that the **Create GCP** button ⊕ looks the same as the Fiducial Locator button in the previous step. The main

thing to remember in the **Multipoint Geometric Correction** window is that the raw image (**schenck.img**) is displayed on the left and the reference image (**schenck_doq.img**) is displayed on the right.

2. You may want to zoom out on both images (⊞ Fit image to window button), so you can figure out where the Schenck Memorial Forest boundary falls in both images. This will make it much easier to place GCPs (Hint: the Schenck is located in the middle of both images).

To place a GCP:

3. Find a feature that can be clearly seen in both images. Features typically used include permanent structures, usually human-made, such as road intersections, parking lots, or building corners. Occasionally natural features, such as a single tree in an open field, can serve as an acceptable ground control.

4. Using the target box for **schenck_doq.img** (the right-hand reference image), zoom in on the selected feature as closely as possible (it will probably fill the close-up mini-viewer window. Then, click on the Create GCP button ⊕ in the **Multipoint Geometric Correction** dialog. Returning to the close-up viewer, click on a selected point to place a GCP on the feature. (You can change the color of the point marker in the GCP Tool dialog.)

   When you do so, geographic coordinates will appear in the dialog under the **X Ref**, **Y Ref**, and **Z Ref** columns (see Figure 4.17).

5. Similar as explained in the earlier point, now use the target box in **schenck.img** (the left-hand image) to zoom in on the feature as much as reasonably possible. Click on the Create GCP button in the GCP Tool dialog. In the close-up viewer for **schenck.img**, place a GCP on the feature, making sure to match as exactly as possible the location you marked on **schenck_doq.img**.

   When you do so, image coordinates will appear in the dialog under **X Input** and **Y Input**... (Figure 4.17).

Orthorectification   111

Point #	Point ID	>	Color	X Input	Y Input	>	Color	X Ref.	Y Ref.	Z Ref.	Type	X Residual	Y Residual
1	GCP #1			2429.663	-6879.759			2079243.270	750916.305	101.226	Control		
2	GCP #2			4544.964	-587.134			2082619.795	753604.614	128.279	Control		
3	GCP #3			5964.446	-4881.160			2081847.790	750769.163	124.609	Control		
4	GCP #4			336.574	-2395.524			2079653.366	754183.770	98.348	Control		
5	GCP #5			3878.588	-3000.902			2081397.528	752557.867	119.716	Control		
6	GCP #6			3590.434	-4887.533			2080548.657	751620.489	119.946	Control		
7	GCP #7	▶				▶					Control		

FIGURE 4.17

Image coordinates under X Input and Y Input columns. (From ERDAS IMAGINE®/Hexagon Geospatial.)

The GCP locations on the unreferenced image (Schenck.img) and the reference image (Schenck_doq.img) are stored in separate files.

6. To save the GCPs, select **File | Save Input As** (do not click on **Save**) from the File menu and save it as **schenck_input2.gcc**. Then choose **File | Save Reference As** from the file menu and save as **schenck_ref2.gcc**.

7. To obtain a RMSE report, ERDAS IMAGINE requires a minimum of six GCPs to be distributed throughout the image. So, you will have to place a minimum of **five more points**, following the above-listed procedures. Try to place the GCPs so that all portions of **schenck. img** get reasonable coverage—this is not always easy because sometimes it is hard to find features that would serve well as ground control points.

Be sure to **save** periodically so that you can get back previously marked points if something happens.

8. After you have placed at least six GCPs on the two images, you can click on the **Solve Geometric Model with Control Points** button ∑ to get an x, y, and total RMSE. These values will be displayed in the lower-right corner of the GCP Tool (Figure 4.18).

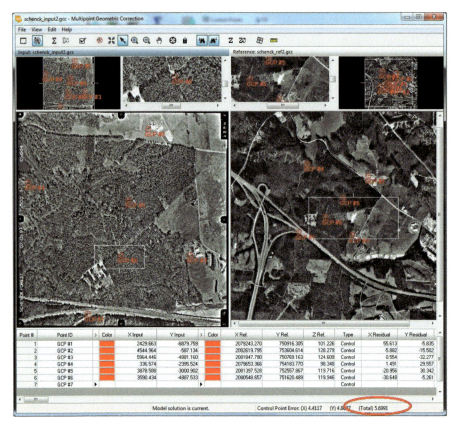

FIGURE 4.18
Total root mean square error (RMSE). (From ERDAS IMAGINE®/Hexagon Geospatial.)

9. In the GCP Tool dialog, the X and Y residuals and RMSE associated with each GCP are displayed, along with the contribution of each point to the total RMSE (Figure 4.19).

X Residual	Y Residual	RMS Error	Contrib
-1.747	4.802	5.110	0.556
-1.256	-3.592	3.806	0.422
-4.469	-4.917	6.644	0.736
14.360	8.227	16.549	1.834
-4.074	-6.954	8.059	0.893
-3.031	-7.474	8.065	0.894

FIGURE 4.19
GCP Tool dialog display with the contribution of each point to the total RMSE. (From ERDAS IMAGINE®/Hexagon Geospatial.)

Orthorectification 113

NOTE: The goal is to reduce the total RMSE to below 10.000 units. If you are still a long way away from this value, you may want to add other GCPs to improve the "solution." (**Be careful:** The GCP Tool may attempt to predict where a GCP will fall in the raw image after you mark it on the DOQQ—however, the location of that automatically placed marker is likely to be incorrect. Always check the location of GCPs yourself; **do not rely on the program!**). You may need to adjust the other GCPs to reduce error.

Once you have enough GCPs, you can use the values in the columns to decide if any of the GCPs are contributing too heavily to the RMSE, and thus should be deleted.

10. Should you need to, you can delete a GCP with the following steps:
    - Highlight the appropriate GCP row in the GCP Tool dialog (it will turn blue).
    - Then, with the cursor under the far left-hand column (the gray column labeled **Point #**), right-click to get a list of options.
    - One of those options will be **Delete Selection**.

    IMPORTANT: Each time you add or remove a GCP, be sure to click on the **Solve Geometric Model…**button $\boxed{\Sigma}$ to re-calculate the RMSE value.

    If all else fails (e.g., if you keep adding and/or deleting GCPs and your RSME does not improve), you can go back and change the positions of the GCP markers on the images. If you click on and highlight one of the GCP rows in the GCP Tool dialog, the target boxes and close-up viewers of both images will move to that point. You can then adjust the position of the markers on either image. It is probably best to move the marker on the uncorrected image (**schenck.img**) rather than the reference image. In any case, do not change the marker locations very much—try to keep the match between images as close as possible, since the markers are supposed to indicate the same geographic location!

11. When you finally get the RMSE below the 10-unit threshold, you can save the final point locations for both the reference (**schenck_doq.img**) and input (**schenck.img**) images. (Select **File | Save Input As** from the File menu, save it as **schenck_input2.gcc**, then choose **File | Save Reference As** from the file menu, and save as **schenck_ref2.gcc**.).

Creating an Orthophoto

Once you have come up with a reasonable RMSE solution, you are ready to generate an orthophoto.

1. Find the Geo Correction Tools toolbar at the top of the GCP editor. Click on the **Display Resample Image Dialog** button (Figure 4.20).

FIGURE 4.20
ERDAS IMAGINE geo correction tools toolbar. (From ERDAS IMAGINE®/Hexagon Geospatial.)

NOTE: This button will not be enabled if you do not have a **RMSE solution** for the image. In other words, you cannot complete this step until you have a minimum of **six GCPs**!

2. The **Resample dialog** will open. There are a number of options you could change here (resample method, cell size, etc.), but leave them on their default settings (Figure 4.21).

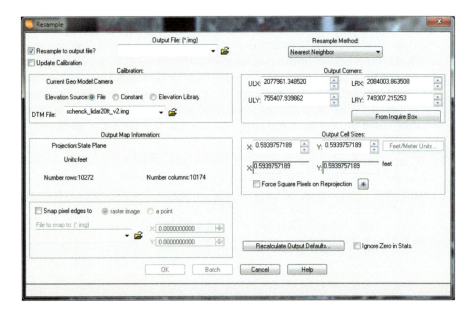

FIGURE 4.21
Resample dialog window. (From ERDAS IMAGINE®/Hexagon Geospatial.)

Orthorectification 115

3. Save the output file as **schenck_ortho.img**. Then click **OK**. A process list bar will appear in the center of the screen. When the status bar reads **Done**, click on **Close** (Figure 4.22).

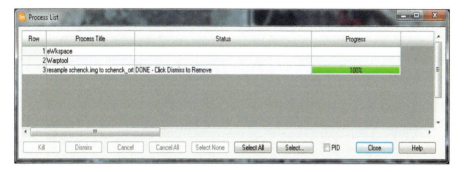

FIGURE 4.22
Process List with completed resample operation. (From ERDAS IMAGINE®/Hexagon Geospatial.)

4. You can close most of the various windows by clicking **Exit** on the **Multipoint Geometric Correction** window. (You may get a prompt to save the recent changes, so go ahead and do so.) You may also right-click in the display viewer (2D View #1) and select "**Clear View**" from the options list to clear the previously loaded **schenck.img**.

Viewing the Orthophoto

You now have a finished orthophoto. You can view the image by opening it in 2D View #1. If you would like to compare the orthophoto to the DOQQ image, do the following:

1. Open **schenck_doq.img** in 2D View #1.
2. Then choose to add another 2D Viewer from the menu ribbon. The 2D View #2 is added to the map display. Add **Schenck_ortho.img** to this view window.
3. You can compare how well the orthophoto and **schenck_doq.img** line up using the **Swipe** tool. Be sure to zoom into various areas throughout the image for comparison. The Swipe tool can be found under the **Home** menu in the View section (Figure 4.23).

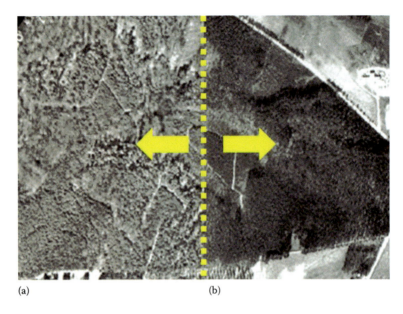

(a) (b)

FIGURE 4.23
Swipe tool comparison of schenck_doq.img (a) and Schenck_ortho.img (b). (From ERDAS IMAGINE®/Hexagon Geospatial.)

Review Questions

1. How is orthorectification different from polynomial transformations and rubber sheeting to georeference an uncorrected digital image to a reference image?
2. The output of orthorectification is an orthophoto, which is described as being *planimetrically accurate*. What does the term planimetrically accurate in an orthophoto refer to?
3. What would be the most important advantage to having a planimetrically accurate image or orthophoto?
4. In creating an orthophoto in a digital image processing package, why is it important that each ground observation, or ground control point (GCP) be selected from features that can be clearly seen in both images?
5. What are the typical attributes of the features used for GCPs?

5

Positional Accuracy Assessment

Overview

Two major accuracy components in remote sensing are referred to as positional accuracy and classification (or thematic) accuracy. Image accuracy assessment has been a key component and the focus of a significant number of remote sensing studies (Congalton 1991, Goodchild et al. 1992, Dai and Khorram 1998, Congalton and Green 1999, Khorram et al. 1999, Paine and Kiser 2003). Without assessing the accuracy of the classified data, the reliability and repeatability of the output products are in question.

Positional accuracy refers to the spatial agreement of an object displayed on an image as compared to the actual position of that object on the ground (true position). Land use/land cover change detection is also both enhanced and challenged by increases in image spatial resolution. A fundamental error source in change analysis is image mis-registration (Dai and Khorram 1998). Co-registration of multi-date high spatial resolution satellite data is limited both by the positional accuracy of image registration reference data, which can be as erroneous as 1 meter (m) even in differentially-corrected Global Positioning System (GPS) data, and by the likelihood that two images captured by a pointed high spatial resolution satellite sensor at different times will have different degrees of topographic distortion.

Positional accuracy, however, should not be confused with the second major type of accuracy in remote sensing, classification accuracy. It is important to note that positional accuracy impacts the thematic accuracy of classified data as well. Classification accuracy does not refer to positional locations of objects with the image to true positions on the ground, but rather the characterization of, typically raw pixel data into categories representing defined land use and land cover groupings or classes, and how well these classes agree with reference data. This reference data typically consists of GPS field-collected points representing true ground positions for each class.

A measure of positional accuracy is the Root Mean Square Error (RMSE). The RMSE (or RMS error) is the absolute fit of the corrected positional accuracy transformation model to the data, or the difference between values predicted by the transformation model and the actual values observed.

117

The RMSE is calculated by computing the difference between each predicted value of the transformation model and the actual observed values, squaring these differences, calculated the sum of the squared differences divided by the total number of values and taking the square root. The result represents the RMSE value (Figure 5.1).

$$\text{RMS error} = \sqrt{\frac{e_1^2 + e_2^2 + e_3^2 + \ldots + e_n^2}{n}}$$

FIGURE 5.1

RMS error equation. Where e represents the difference between each predicted value of the transformation model and the actual observed values, and n represents the total number of values.

The Total RMSE refers the total root mean square (RMS) error of the whole image. The Total RMSE incorporates individual RMS errors for each observation or ground control point (GCP). In general terms, the RMSE is a measure of how much distortion, or stretching, has occurred in the image (in the X or Y directions, or both) that is being georeferenced. The farther out of shape the image becomes, the larger the RMS.

Every image is potentially distorted in the X or Y direction, or both. In other words, the coordinate space of an image is a projection. If the *projection* of the image is compatible with the projection of the data that is being used as a referencing points (such as actual ground positions) then the RMS is dependent on how closely the *From points* in the image relate to the *To points* in the data (such as actual ground positions). The minimum RMS is determined by the resolution of the image (and your computer screen). The RMSE can also be viewed as the error occurring between the source and destination control points. For each transformation performed, a root mean square error will be calculated to provide an indication of how well the transformation fits the data.

Another measure useful in the determination of positional accuracy is the Euclidean distance. The Euclidean distance is also referred to as the *straight-line distance*. This distance is computed based on the equation of a straight line. Mathematically, the Euclidean distance between two points can be found using the distance formula. The distance between (x_1, y_1) and (x_2, y_2) is given by Figure 5.2:

$$d = \sqrt{(x_2 - x_1)^2 + (y_2 - y_1)^2}$$

FIGURE 5.2

Euclidean distance between two points can be found using the distance formula. Where x represents the X-position and y represents the Y-position in coordinate space represents straight line distance.

Positional Accuracy Assessment 119

Determining how far away the sample or ground control points on an image (output image [i]) are away from the measured or known position (reference image [c]) on the surface of the Earth is known as positional accuracy. This exercise will determine the positional accuracy of an image using the MS Excel worksheet (*.xls format) provided in the companion media. On the worksheet, complete the table (columns E, F, G, H, and I). Then use these calculations to calculate the SUM, N, and RMSE for columns F and H. Finally, calculate the Average Euclidean Distance from column I.

Positional Accuracy Application

Learning Objectives

1. To calculate positional accuracy
2. To understand the relationship between the output image (i) and the reference image (c)

Data Required: Positional Accuracy Worksheet.xls

The data provided for this exercise was derived from actual project data collected within the Department of Forestry and Natural Resources and adapted from original research developed by Dr. Heather Cheshire at North Carolina State University. The X and Y control coordinates came from field measurements using a high accuracy GPS unit. As a result, this data represents an independent dataset not used in the orthorectification of the image. The reference data for the orthorectification of the previous exercise was in fact, another orthorectified image and the RMSE for this process was < 0.6096 m (< 2 feet).

Calculating Error in the X and Y Directions

The following equation will be used to determine the RMSE and Euclidean distance provided on the accompanying MS Excel worksheet. This data represents actual image derived points and field-collected GPS coordinates that were used to rectify the image data.

The equations you will use in this exercise as follows:

$$X_i - X_c$$

where:
X_i represents points in output image (the *rectified* image product)
X_c represents points from reference image, or from GPS coordinates, and so on. (assumed to be correct)

$$Y_i - Y_c$$

where:
Y_i represents points in output image (the *rectified* image product)
Y_c represents points from reference image, or from GPS coordinates, and so on. (assumed to be correct)

Root Mean Square Error

RMSE is used to determine how well your transformation model solution fits the original data. These equations indicate the difference between the observation and the calculated value for each observation within the data (Figure 5.3).

$$\text{RMS error} = \sqrt{\frac{X_1^2 + X_2^2 + X_3^2 + \dots + X_n^2}{n}}$$

$$\text{RMS error} = \sqrt{\frac{Y_1^2 + Y_2^2 + Y_3^2 + \dots + Y_n^2}{n}}$$

FIGURE 5.3
RMS error equations for X and Y positions in coordinate space.

Total Root Mean Square Error

Total RMS error represents the total RMS error for a complete image, or the error occurring between the source and the destination control points. The Total RMS error can also be used as an overall indication of how acceptable the derived transformation is. Please note that an acceptable level of RMS error may be quite subjective, depending on the intended use and goal of the transformed data.

Positional Accuracy Assessment

Euclidean Distance

Euclidean distance, or the straight-line distance, may be calculated using the Euclidean distance formula. See the Euclidean distance formula given earlier for calculation of the distance between (x_1, y_1) and (x_2, y_2) (Figure 5.2).

Review Questions

1. How is positional accuracy different from classification accuracy?
2. True or False: Every image is potentially distorted in the X or Y direction, or both?
3. Define the term Root Mean Square Error (RMSE). What does this term represent?
4. How is the RMSE value calculated?
5. The Total RMSE refers the total root mean square (RMS) error for an entire image. The Total RMSE takes into account the individual RMS errors for each observation or ground control point (GCP). In general terms, how does the RMSE relate to the transformation within the image?

6

Radiometric Image Enhancement

Overview

Typically satellite imagery is often captured from hundreds of miles away from Earth, within a specified orbit. This distance, coupled with the radiometric resolution of the sensor and the band availability, provides important information necessary to understand the resolving capability expected from any data output product acquired from the sensor. The radiometric resolution refers to the dynamic range, or a number of tonal levels, at which data for a given spectral band are recorded by a particular sensor. The distance from Earth, coupled with the radiometric resolution of the sensor provides important information necessary to understand the resolving capability expected from the sensor. Often, without the application of radiometric enhancement techniques, the raw data captured by the sensor is too dark to be useful by the image analyst in interpreting features on the ground. Spectral image enhancement techniques included the band combination creation of true color composite images and false color composite images (i.e., vegetation/forested areas displayed as varying shades of red).

Radiometric enhancements adjust the contrast, or in some cases, stretch the tonal levels of the radiometric range of the data. For example, an image with a 1-bit image radiometric resolution would contain two brightness values, or grayscale levels (black and white), within the data. An image with an 8-bit radiometric resolution would have up to 256 grayscale levels within the data. In contrast, an image with an 11-bit image may have the capability of containing up to 2,048 grayscale levels (Figure 5.4). Landsat 4–7 possess an 8-bit radiometric resolution. The Landsat 8 sensor possesses a radiometric resolution captured in a 12-bit dynamic range (4,096 grayscale levels) and delivered as 16-bit images when processed into Level-1 data products (scaled to 55,000 grayscale levels). The 12-bit dynamic range radiometric resolution of the Landsat 8 sensors if also comparable to Sentinel-2's radiometric range (Figure 6.1).

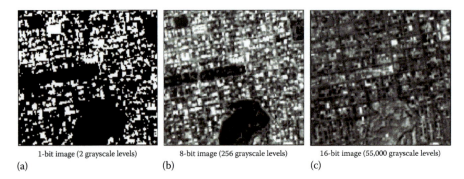

| 1-bit image (2 grayscale levels) | 8-bit image (256 grayscale levels) | 16-bit image (55,000 grayscale levels) |
| (a) | (b) | (c) |

FIGURE 6.1
The concept of radiometric resolution is illustrated by three images. (a) The image represents a 1-bit image, where two brightness values, or grayscale levels (black and white), are portrayed within the data. (b) The image represents an 8-bit image, where up to 256 grayscale levels are portrayed within the data. (c) The image represents a 16-bit image, where up to 55,000 grayscale levels are portrayed within the data.

Applying radiometric enhancements techniques may improve the contrast between features within the image being represented within a specific radiometric range, pixel tonal level, or grayscale values. However, caution should be exercised in applying contrast enhancements or stretches in middle values as too large an adjustment may not differentiate the lower and higher end values of the data's radiometric range.

Radiometric Enhancement Application

Learning Objectives

1. To demonstrate the basic understanding of one of the two major image enhancement techniques:
   - Radiometric enhancement: An image enhancement process that is used to adjust the original pixel values within an image based on a function that is used to improve the interpretability of the output (screen) pixel values.
   - Spatial enhancements: An image enhancement process that is used to adjust the original pixel values based on the surrounding pixel values or neighborhood of pixel values.
2. To perform radiometric enhancements in an effort to produce a more interpretable image from the raw image.

Data Required: This exercise will focus on the Raleigh, NC, USA region of the Landsat 5 TM image **LT05_L1TP_016035_19941220_20160926_01_T1**, Acquisition Date: 1994/12/20. This scene represents an image captured by the Landsat 5 TM sensor. Components of this exercise adapted from

Radiometric Image Enhancement 125

University of Oregon on-line resources available at: http://geog.uoregon.edu/amarcus/geog418w07/index.html.

Performing Radiometric Enhancements

Loading Stretched and Non-Stretched Images

1. Download the image **LT05_L1TP_016035_19941220_20160926_01_T1** from the EarthExplorer website (http://earthexplorer.usgs.gov/) following the procedure demonstrated in Chapter 1. Perform the Layer Stack operation in ERDAS IMAGINE® using bands 1–5 (see Table 2.2).

2. In ERDAS IMAGINE, open the newly created multi-band image, **LT05_L1TP_016035_19941220_20160926_01_T1_ALB.img**, using the True Color Composite (TCC) band combination (**Red 3, Green 2, Blue 1**) as completed in a previous exercise in Chapter 2.

    - In the "**Select Layer To Add**" dialog, highlight the **LT05_L1TP_016035_19941220_20160926_01_T1_ALB.img**
    - Select the "**Raster Options**" tab and ensure the band selections **Red 3, Green 2, and Blue 1**.
    - Now select **OK** (Figure 6.2).

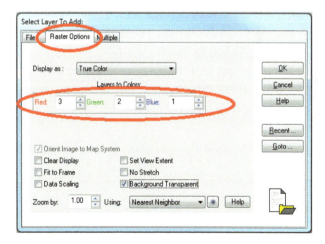

FIGURE 6.2
Raster band combination. (From ERDAS IMAGINE®/Hexagon Geospatial.)

3. Zoom into the Raleigh, North Carolina region of the image as seen in the following image (Figure 6.3).

FIGURE 6.3
Raleigh North Carolina region within the full data scene. (From ERDAS IMAGINE®/Hexagon Geospatial.)

4. Open a second image in a new 2D View window with no stretch applied.

- From the "**Home**" menu tab, select the "**Add Views**" button to open a second viewer (**Create New 2D View**).
- Open a second version of the **LT05_L1TP_016035_19941220_20160926_01_T1_ALB.img** in 2D View #2.
- Apply band combination **Red 3, Green 2,** and **Blue 1**.
- This time check the "**No Stretch**" box on the **Raster options** tab.
- Now select **OK** (Figure 6.4).

Radiometric Image Enhancement 127

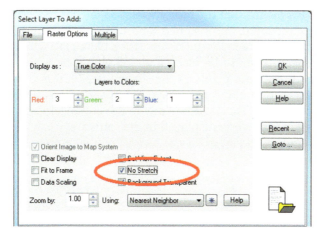

FIGURE 6.4
No stretch raster option. (From ERDAS IMAGINE®/Hexagon Geospatial.)

NOTE: With the same image displayed in two separate 2D View windows, it may be helpful to have both images zoomed into the same area (Raleigh, North Carolina area) by linking the open views (select the **Link Views** icon) as demonstrated in Chapter 2 "Linking Images in Graphics Windows," as well as using the **Zoom** tool.

Understanding the Stretch

To create an appropriate radiometric stretch of an image to target the interpretation of a specific feature, the raw data values that are associated with the feature of interest should be identified to determine the range of these values within the image histrogram(s).

The image in the first viewer (2D View #1) uses the ERDAS default setting of +2 standard deviations from the mean of the file values. This applies a standard deviation stretch (2.0) that allows you to see the features more clearly.

The image in the second viewer (2D View #2) loads the image without any stretching (a standard deviation stretch of 0).

5. Using the **Metadata** tool look at the general information and histograms for each band of the image in the first viewer (2D View #1).

   Pay close attention to "**Data Type**" in the Layer Info section and "**Mean**" of the data pixels in the Statistics Info section. The "**Data Type**" (bit depth) refers to the radiometric range of values that each cell within the image contains. This image is specified as "Unsigned

8-bit." This means that the radiometric range goes from 0 to 255. All things being equal, an unsigned 8-bit raster would produce greater image detail than a lower-bit depth raster. On a gray scale range, this would mean 0 would be black and 255 would be white. Since the "**Mean**" of the data pixels in this image is 37.8, this indicates that more pixels throughout the image are closer to 0 (black) than they are to 255 (white). Thus, without any image enhancements, the image will appear dark when displayed in the software.

Close the Metadata window when done (Figure 6.5).

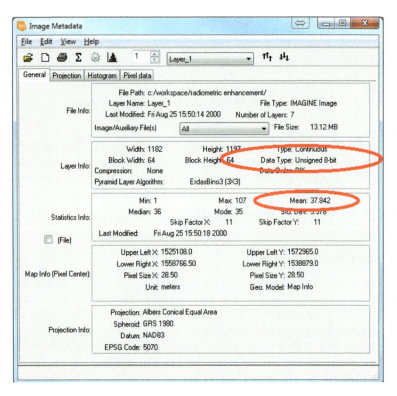

FIGURE 6.5
Image metadata window. (From ERDAS IMAGINE®/Hexagon Geospatial.)

Radiometric Image Enhancement 129

6. Using the "**Inquire**" cursor, explore the file pixel values for multiple locations throughout the first image (2D View #1). Notice differences in the File Pixel values (Input Values) and Lookup Table Values (LUT values) for bands **Blue 1, Green 2,** and **Red 3** as you move the cursor across different features within the image (Figure 6.6).

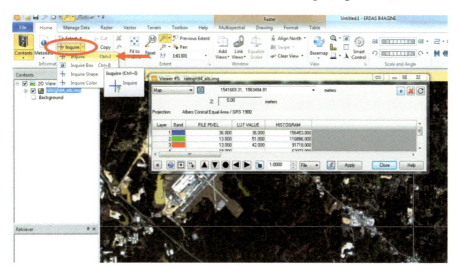

FIGURE 6.6
Inquire cursor. (From ERDAS IMAGINE®/Hexagon Geospatial.)

ERDAS IMAGINE and Lookup Tables (LUT Values)

ERDAS IMAGINE and most image processing applications use a defined set of values to enhance contrast, or make the displayed image appear brighter on the computer screen. These values, usually organized in tables to make it easier to interpret or lookup the transformations being applied to the data, are known as lookup tables. Lookup tables (LUT) contain values that the software uses to adjust the original file pixel values (usually lower values, i.e., darker, lower contrast, etc.) into higher values (i.e., brighter) or a wider range of values (i.e., higher contrast) for display on the screen or to print an image.

TABLE 6.1

Comparing the Raw Image Pixel File Values to a Translated Lookup Table (LUT) in the ERDAS IMAGINE Software

Input Value		LUT Value
30	→	0
31	→	25
32	→	51
33	→	76
34	→	102
35	→	127
36	→	153
37	→	178
38	→	204
39	→	229
40	→	255

Source: Intergraph Corporation, *ERDAS Field Guide*, Intergraph Corporation, Huntsville, AL, 792 p, 2013.

For example, the lookup table below transforms the original file pixel values (input), that exhibit a narrow range of values that are too close together to produce an image with enough contrast or brightness for general interpretation. Each value in the input image is assigned to new values that widen, or stretch, the range from 0 to 255 (Table 6.1).

Close the Inquire cursor tool window when done.

7. Using the "**Spectral Profile**" tool , collect spectral data for several water points and several forested points (**Multispectral | Spectral Profile**).

In the "**Spectral Profile #1 --> Viewer #1**" tool window that opens, select the point tool button ➕ **Create New Profile Point in Viewer** on the menu to select a point of interest (Figure 6.7).

Radiometric Image Enhancement 131

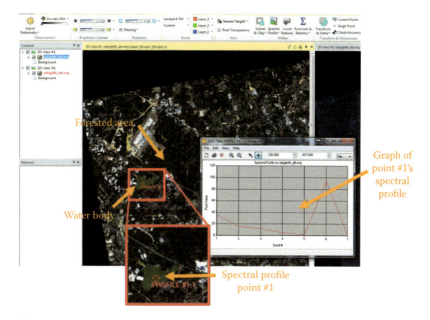

FIGURE 6.7
Creation of new profile point in profile graph viewer. (From ERDAS IMAGINE®/Hexagon Geospatial.)

8. Use the "**View | Tabular Data**" option on the Spectral Profile tool window to review the values for the layers/bands (be sure to also pay attention to the values for band 4 specifically) at each point.

   NOTE: After sampling several different water and forest locations throughout the image, note roughly, the average minimum and maximum values for the water features and the forest feature in band 4 (use them below with the breakpoint manipulations in the "Adjusting the Stretch" section that follows).

9. Also examine if the histogram for each band within the image is unimodal or bimodal by going into "**Metadata**" on the Home tab and looking at the Histogram tab (again, you may want to closely examine band 4).

   Remember, the histogram's plot will give you the image layer histogram (e.g., Layer 1 = Band 1, Layer 2 = Band 2, Layer 3 = Band 3, etc.) for the whole image and not each profile point that you create using the Spectral Profile tool.

A histogram is considered **unimodal** if there is one hump as seen in Figure 6.8.

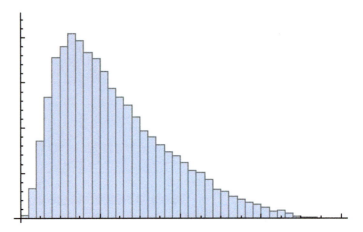

FIGURE 6.8
Unimodal histogram distribution.

A histogram is considered **bimodal** if there are two humps (in some cases multimodal if there are more than two) (Figure 6.9).

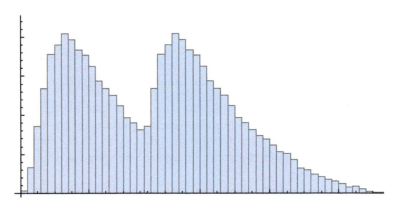

FIGURE 6.9
Bimodal histogram distribution.

Radiometric Image Enhancement 133

Adjusting the Stretch

Image enhance approaches in ERDAS IMAGINE can be accessed very generally by using simple *dial-like* controls for brightness and contrast adjustments.

1. Adjust the brightness and contrast of your image by selecting the **Brightness Contrast** options from the "**Multispectral**" menu tab (Figure 6.10).

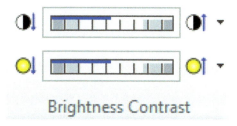

FIGURE 6.10
Brightness contrast adjustment options. (From ERDAS IMAGINE®/Hexagon Geospatial.)

Adjust the brightness and contrast dials and examine the change each adjustment makes to the image. The image can be reset to the original or default condition by selecting the **Reset** options from the drop down tabs associated with each adjustment dial (Figure 6.11).

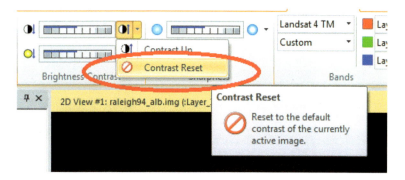

FIGURE 6.11
Brightness contrast options reset. (From ERDAS IMAGINE®/Hexagon Geospatial.)

2. Now select the **Adjust Radiometry** button from the **Enhancement** grouping on the **Multispectral** tab and select "**General Contrast**."

NOTE: In earlier versions of ERDAS IMAGINE (such as versions 2010–2013) the **Adjust Radiometry** button on the Multispectral menu tab will be labeled as the **General Contrast** button (Figure 6.12).

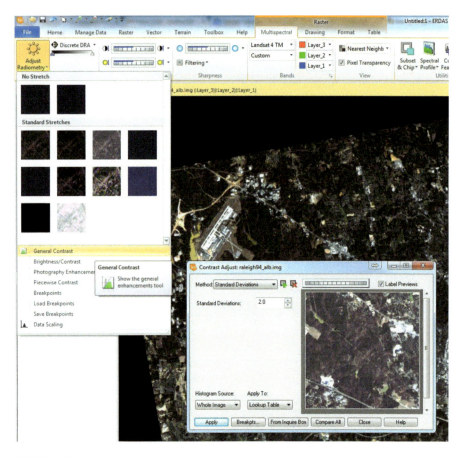

FIGURE 6.12
Contrast adjust options. (From ERDAS IMAGINE®/Hexagon Geospatial.)

3. In the "**Contrast Adjust**" window, use the "**Method**" drop-down selections to examine different methods of enhancements available (Figure 6.13).

Radiometric Image Enhancement 135

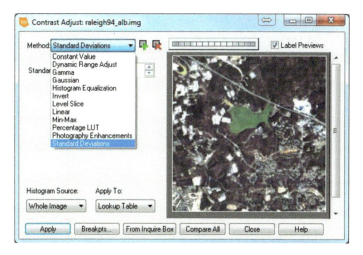

FIGURE 6.13
Contrast adjustment methods. (From ERDAS IMAGINE®/Hexagon Geospatial.)

4. By selecting the **Apply** button, you can commit all changes to the image. Experiment with applying several of the different methods available to see how the image changes in the preview window.
5. Next, in the Contrast Adjust window, select the "**Breakpts…**" button on the bottom row of buttons within the dialog window to open the Breakpoint Editor (Figure 6.14):

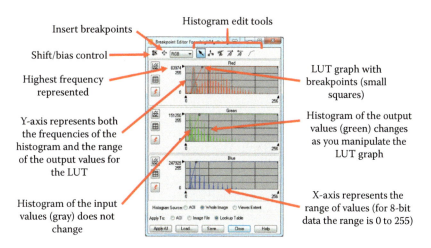

FIGURE 6.14
Breakpoint editor. (From ERDAS IMAGINE®/Hexagon Geospatial; Adapted from ERDAS, 2001, ERDAS IMAGINE Tour Guides, ERDAS IMAGINE V 8.5. ERDAS Inc., Atlanta, GA, p. 31.)

The graph that follows demonstrates how original pixel values are increased to enhance displayed image display. For example an original pixel value of 32 (darker) is adjusted to 68 (brighter) and an output screen value of 129, resulting in a brighter display of the features that were in the original 32 (and above) pixel value range (Figure 6.15).

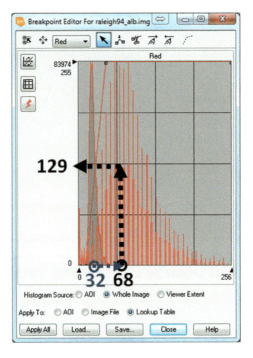

FIGURE 6.15
Breakpoint and pixel display histogram for a single band (red). (From ERDAS IMAGINE®/Hexagon Geospatial.)

In the figure above, the darker (gray) histogram represents the original image pixel values, ranging from 0 (black) to 255 (white) along the x-axis of the graph. The y-axis represents the output screen values of the displayed image. The pixel value of 54 that is located more towards the darker end of the 0–225 range would be adjusted as a result of the stretch to an output pixel value of 148 which represents a brighter value on the screen. Additionally, pixel values that were originally grouped in the gray histogram more towards the darker end of the graph (i.e., closer to 0) are now stretched somewhat more evenly between 0 and 255 (i.e., red histogram).

Radiometric Image Enhancement   137

Making Finer Adjustments

1. Now remove all layers from the viewer and close viewer 2 (2D View #2) so that you have only one viewer open (2D View #1).
2. Load the **LT05_L1TP_016035_19941220_20160926_01_T1_ALB.img** image as a gray scale image (**Raster Options | Display as: Gray Scale**) and select band 4 (**Raster Options | Layer: 4**) (Figure 6.16).

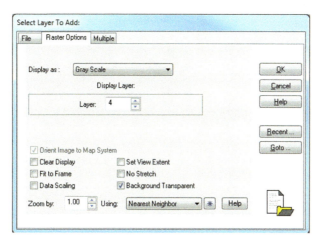

FIGURE 6.16
Adding a gray scale image. (From ERDAS IMAGINE®/Hexagon Geospatial.)

3. Select "**OK**" to load the gray scale image.
4. Re-open the "**Breakpoint Editor**" window from the "**Contrast Adjust**" window (**Raster** menu tab group | **Panchromatic** tab | **Adjust Radiometry** (Enhancement grouping) | **General Contrast**) (Figure 6.17).

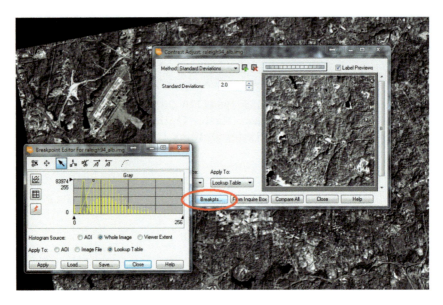

FIGURE 6.17
Standard deviation adjustment for the gray scale image. (From ERDAS IMAGINE®/Hexagon Geospatial.)

5. Change the Standard Deviations to 1.0 in the "**Contrast Adjust**" window and click on "**Apply**."
6. Change the Standard Deviations back to the default (2.0).
7. Next, experiment with adjusting the breakpoints.

   NOTE: You can click on and drag the existing breakpoints, or use "**Shift-click on**" (hold down the shift button on the keyboard and left-click with the mouse) in the Breakpoint Editor window and insert one (or more) breakpoint(s). Try dragging the breakpoints or inserting some new breakpoints ("Shift-Click" in the desired location), then examine how the image changes. Remember, you must click on "**Apply**" to see how this affects the image.

   For example, examine the following three images (Figures 6.18 through 6.20).

Radiometric Image Enhancement 139

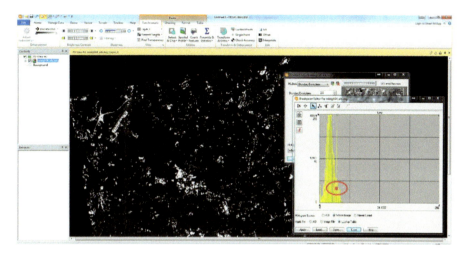

FIGURE 6.18
Breakpoint adjustment. (From ERDAS IMAGINE®/Hexagon Geospatial.)

In the earlier image, "Shift-Click" was used to insert a single breakpoint and move the histogram of the output LUT values (yellow) to match the image input file/pixel values (gray). Notice that after clicking on "Apply," the image generally darkens and is similar to loading the original image with the "No Stretch" option checked (a standard deviation of 0).

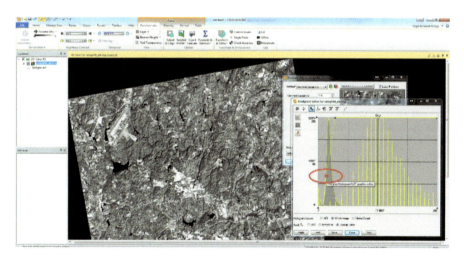

FIGURE 6.19
Breakpoint adjustments to stretch the histogram of the output LUT values (yellow) across the x-axis to allow the image to appear similar to the default stretch (2.0 standard deviation). (From ERDAS IMAGINE®/Hexagon Geospatial.)

In the previous image the single inserted breakpoint was used to stretch the histogram of the output LUT values (yellow) across the x-axis. Notice that after clicking on "**Apply**" the image appears similar to loading the original image with the default stretch (2.0 standard deviation).

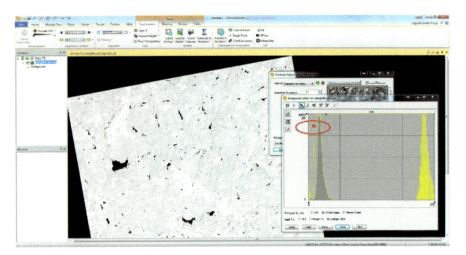

FIGURE 6.20
Breakpoint adjustments to stretch the histogram of the output LUT values (yellow) across the x-axis to allow the image to appear brighter than the image displayed with a default stretch. (From ERDAS IMAGINE®/Hexagon Geospatial.)

In the earlier image the single inserted breakpoint was used to move the histogram of the output LUT values (yellow) to the right-end of the x-axis. Notice that after clicking on "**Apply**" the image appears to brighten, similar to previous adjustments made with the **Brightness Contrast** tool.

You can also insert breakpoints to highlight a certain feature you are interested in, especially if you have noted the range of spectral values the feature encompasses in the image (e.g., by using Spectral Profile tool).

8. Click on the "**Start table editor for gray**" icon in the Breakpoints Editor window to open the "**Gray Lookup Table**" dialog window (at this point, if you would like to start fresh by reloading the grayscale image, you may do so, and then open the "**Gray Lookup Table**") (Figure 6.21).

Radiometric Image Enhancement 141

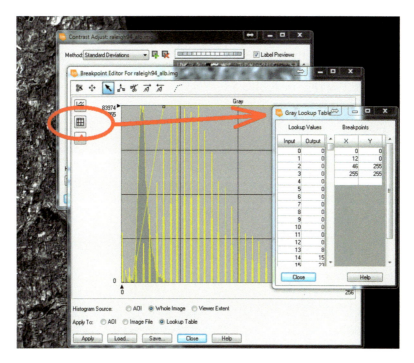

FIGURE 6.21
Lookup table editor. (From ERDAS IMAGINE®/Hexagon Geospatial.)

9. Review the X and Y columns in the Gray Lookup Table. The initial image had breakpoints (X values) at 0, 12, 46, and 255 (Figure 6.22).

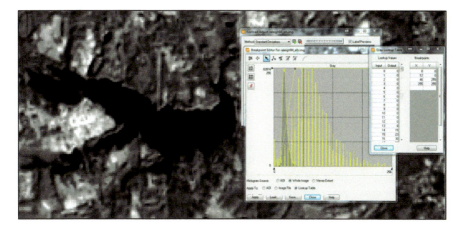

FIGURE 6.22
Gray lookup table displaying original image breakpoints (X values) at 0, 12, 46, and 255. (From ERDAS IMAGINE®/Hexagon Geospatial.)

142   *Image Processing and Data Analysis with ERDAS IMAGINE®*

The following image on the left shows the initial image's breakpoints values (highlighted in red) at: 0, 0; 12, 0; 46, 255; and 255, 255. The image on the right shows a highlight of the gray histogram where the majority of the range of image data occurs (between 12, 0 and 46, 255) (Figure 6.23).

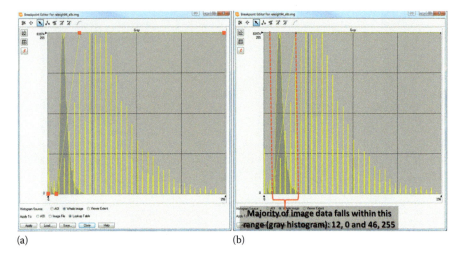

(a)   (b)

FIGURE 6.23
(a) The graph represents the original image breakpoints (X values). (b) The graph represents the range of the majority of image data. (From ERDAS IMAGINE®/Hexagon Geospatial.)

Let us say the water values (samples with the aforementioned Spectral Profile tool) ranged from 3 to 5. The stretched values should be a little wider; roughly between 1 and 10. This wider range will help ensure and variations are highlighted that may be missed if the range is set too narrow. Set all other breakpoints to 0, 0 or 255, 255 (Figure 6.24).

Radiometric Image Enhancement

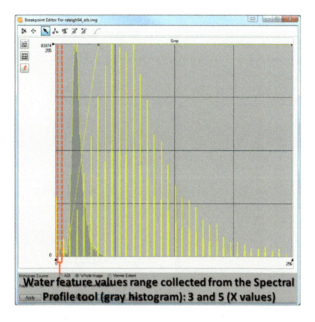

FIGURE 6.24
The dotted red lines represent the range where the water features occur, as collected from the Spectral Profile tool. (From ERDAS IMAGINE®/Hexagon Geospatial.)

10. Using the "**Gray Lookup Table**" window, click within the Breakpoint column cells to add the following X, Y values:

   0, 0 to provide the lower end of possible values

   1, 0 to set all pixels with a digital number (DN) less than 1 to 0 (black)

   10, 255 to set all values greater than 10 to 255 (white)

   255, 255 to set the upper end of possible DN values.

   Applying these options will stretch all pixels with DN between 1 and 10 (water) and make all others either white or black (Figure 6.25).

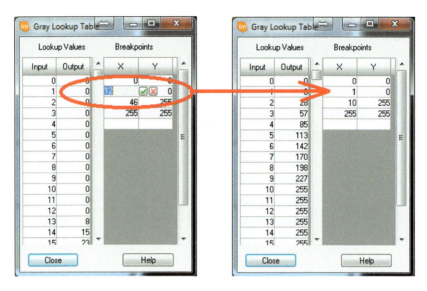

FIGURE 6.25
Lookup table adjustments to highlight water features. (From ERDAS IMAGINE®/Hexagon Geospatial.)

11. Click on "**Apply**" in the **Breakpoint Editor** dialog window to see how the changes affect the image (Figure 6.26).

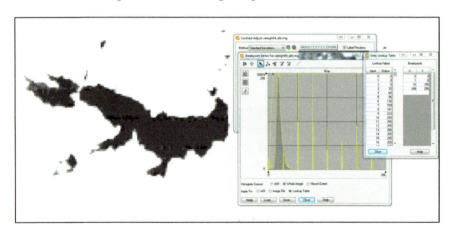

FIGURE 6.26
Image display with breakpoint adjustments highlighting water features. (From ERDAS IMAGINE®/Hexagon Geospatial.)

The following images compare the original image with the standard deviation 2.0 stretch (left) and the new image created to highlight the variations in the water values (right) from the range sampled using the Spectral Profile tool (Figure 6.27).

Radiometric Image Enhancement 145

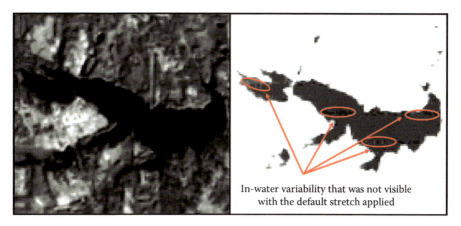

FIGURE 6.27
Comparison of original grayscale image (left) with image adjusted to highlight water features (right). (From ERDAS IMAGINE®/Hexagon Geospatial.)

12. Zoom into Lake Wheeler in the southwestern portion of the image (the following image shows the location of Lake Wheeler, displayed as a TCC image) and inspect these changes closely for yourself (Figure 6.28).

FIGURE 6.28
Location of Lake Wheeler in the image displayed as a TCC. (From ERDAS IMAGINE®/Hexagon Geospatial.)

146   *Image Processing and Data Analysis with ERDAS IMAGINE®*

Applying radiometric stretches to an image represent an effective means of highlighted features of interest within your imagery for further examination. By identifying the specific range the pixels occupy that represent your feature of interest, you are able to adjust the stretch to allow this feature to become more interpretable.

Review Questions

1. Define radiometric resolution as it relates to data for a given spectral band of a given sensor.

2. Why is it necessary, in most digital image processing packages, to apply radiometric enhancement techniques to the raw imagery being displayed?

3. How do radiometric enhancement techniques work to increase the interpretability of the raw data displayed in an image?

4. What is the radiometric range, or how many tonal levels of gray are within data that possesses a 1-bit, 8-bit, 11-bit, 12-bit (dynamic), and a 16-bit range?

5. Assuming an unsigned 8-bit radiometric resolution image (256 grayscale levels) composed of three land cover classes, (1-clear, deep, water/lake, 2-homogenous deciduous vegetation, 3-homogenous, imperious surface) which land cover class would be expected to produce the darkest pixel values?

6. In the theoretical image described in the Question 5, which land cover class would be expected to produce the brightest pixel values?

7

Spatial Image Enhancement

Overview

The previous chapter explored many different techniques image analysts use to radiometrically enhance raw digital images. The techniques learned go way beyond the default Standard Deviation stretch applied to data when loaded in many of the image processing packages to enhance features of interest within the image data. Spatial enhancements, like radiometric enhancements, also improve the interpretability of features within the data. However, instead of operating on individual pixels, as in the radiometric enhancement models, the spatial enhancement models operate by modifying neighborhood pixel values based on the value of a targeted pixel. Common spatial enhancement operations include the following; convolution filter, non-directional edge detection, focal analysis, statistical filters, and image degrade.

The convolution filter applies a matrix of small neighborhoods of cells (e.g., 3×3, 5×5, 9×9, etc.) to compute a neighborhood average as the matrix moves throughout the entire image. The result, in the case of a low pass convolution filter, is that the image may appear smoother, or blurred (less detailed), in appearance based on the filter size. Low pass filters are spatial enhancements that decrease differences between the target pixel and its surrounding pixels. These filters can be useful for removing image noise and variability.

High pass filters are spatial enhancements that increase differences between the targeted pixel and its surrounding pixels. A high pass convolution filter may highlight the boundaries, or edges, occurring between homogeneous groups of pixels.

The non-directional edge filter may perform similar to the high pass convolution filter in highlighting boundaries, or edges, occurring between homogeneous groups of pixels. However, results are derived from the average of two orthogonal 1st derivative edge detectors. The results of this filter may provide a sharp contrast between boundaries on non-linear features.

The focal analysis filter also performs similar to the convolution filter operations. This filter operates on class values within an image file

147

to reduce variability. The statistical filter averages the pixel values using a moving window operation within a defined statistical range.

The image degrade operation can be used to increase the image spatial pixel values to a specified size (from 10 m to 30 m, etc.). Additional detail on the spatial image enhancement operations can be found in the ERDAS IMAGINE® Online Help documentation, the ERDAS IMAGINE Field Guide (Intergraph Corporation. 2013. *ERDAS Field Guide*. Huntsville, AL. 792 p.).

Spatial Image Enhancement Application

Learning Objectives

1. To demonstrate the basic understanding of one of the two major image enhancement techniques:

   - Radiometric enhancement: An image enhancement process that is used to adjust the original pixel values within an image based on a function that is used to improve the interpretability of the output (screen) pixel values.

   - Spatial enhancements: An image enhancement process that is used to adjust the original pixel values based on the surrounding pixel values or neighborhood.

2. To perform spatial enhancements in an effort to make the supplied raw imagery more interpretable.

Data Required: **LT05_L1TP_016035_19941220_20160926_01_T1_ALB.img** (1994 Landsat 5 TM). This scene represents an image captured by the Landsat 5 TM sensor. Components of this exercise adapted from University of Oregon on-line resources available at: http://geog.uoregon.edu/amarcus/geog418w07/index.html.

Spatial Image Enhancements

The following exercise will apply a few standard spatial enhancements, or filters. It can also be helpful to think of spatial filters in three broad categories mentioned previously:

- **Low pass filters:** emphasize low frequency to smooth out image noise or reduce spikes in the data.

- **High pass filters:** emphasize high-frequency to sharpen or enhance linear features such as roads, fault lines, and land/water boundaries.

Spatial Image Enhancement

- **Edge detection filters:** emphasize edges surrounding objects or features in an image to make them easier to analyze.
    1. Open the **LT05_L1TP_016035_19941220_20160926_01_T1_ALB.img** data in viewer window and zoom into the Raleigh, North Carolina area as you did in the previous chapter.
    2. Open the Convolution filter tool dialog window: From the file menu, click on the **Raster menu** tab | **Spatial** in Resolution group, then select **Convolution** and Click on the **Help** button to review this tool's information (Figure 7.1).

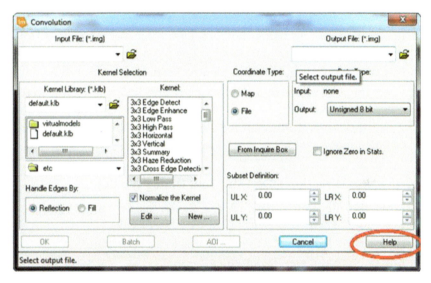

FIGURE 7.1
Convolution filter tool dialog window. (From ERDAS IMAGINE®/Hexagon Geospatial.)

Initiate a 3×3 Filter

3. Enter **LT05_L1TP_016035_19941220_20160926_01_T1_ALB.img** as the input file.
4. Enter **raleigh_3×3 edge** as the output file.
5. Chose the "**3×3 Edge Detect**" from the list of available kernels and click OK.
6. Open the new image in a second viewer and link the two viewers together as accomplished in the ERDAS IMAGINE introductory exercise.
7. Zoom into several different locations and compare these locations on each image.

150　　*Image Processing and Data Analysis with ERDAS IMAGINE®*

Initiate a 5×5 High Pass Filter

8. Repeat this process using the "**5×5 High Pass**" kernel created from the original data.
9. Name the newly created data **raleigh_5×5hipass** (output file).
10. Open the new image in a third viewer and link this viewer to the other two viewers.
11. Zoom into several different locations and compare these locations on each image.

Initiate a 5×5 Low Pass Filter

12. Repeat this process using the "**5×5 Low Pass**" kernel created from the original data.
13. Name the newly created data **raleigh_5×5lopass** (output file).
14. Open the new image in a fourth viewer and link this viewer to the other three viewers.
15. Zoom into several different locations and compare these locations on each image.

Open the Non-Direction Edge Filter Tool Dialog Window

16. From the file menu, click on the **Raster menu** tab | **Spatial** in Resolution group, then select **Non-direction Edge**.
17. Click on the **help** button to review this tool's information.
18. Initiate the **Non-direction Edge filter** tool.
19. Enter **LT05_L1TP_016035_19941220_20160926_01_T1_ALB.img** as the input file.
20. Enter **raleigh_nonedge** as the output file.
21. Under "**Output Options**" select the "**Prewitt**" option and click **OK**.
22. Open the new image in a 5th viewer.
23. Zoom into several different locations and compare these locations with the original data.

Spatial Image Enhancement 151

Open the Image Degrade Tool Dialog Window

24. From the file menu, click on the **Raster menu** tab | **Spatial** in Resolution group, then select **Degrade**.
25. Click on the **help** button to review this tool's information.
26. Initiate the **Image Degrade** tool.
27. Enter **LT05_L1TP_016035_19941220_20160926_01_T1_ALB.img** as the input file.
28. Enter **raleigh_degrade** as the output file.
29. Enter an X and a Y scaling factor of "**5**" by which to degrade the image and click **OK**.
30. Open the new image in a 6th viewer.
31. Finally, zoom into several different locations and compare these locations with the original data.

Review Questions

1. Spatial enhancements, like radiometric enhancements, also improve the interpretability of features within the data. How do spatial enhancement techniques differ from radiometric enhancements?
2. Name the most common spatial enhancement operations.
3. Name the most common spectral image enhancement techniques.
4. How is the convolution filter used to increase image interpretability?
5. Explain the difference between how low pass and high pass filters work to improve image interpretability.

8

Image Digitizing and Interpretation

Overview

Digitization is a method of data input that has traditionally required a manual procedure to enter spatial data into a digital database from physical or hard photographs, maps, or other imagery. This method, while very laborious and time-consuming, could also introduce several areas associated with the data import process and the creation of the database topology. Often the database created from manual digitization is stored as vector information. This vector information can then be used for land use and land cover classification or other land cover analysis operations. Modern Geographic Information Systems (GIS) and image processing software packages now allow for a *heads-up* digitization method. This method allows the analyst to digitize features from photographs, maps, or other imagery directly from a computer screen using a mouse cursor as the input device. Each mouse-click inserts nodes or polygon vertices into the captured vector database. Additionally, modern software also allows for a semi-automated feature extraction technique. This technique allows for faster recognition of boundary features within the image data being digitized. This semi-automated feature extraction technique greatly reduces the amount of labor, times, and errors associated with traditional digitizing.

Image Digitization and Interpretation Application

This exercise will explore the classical (traditional) digitizing approach with help from the ERDAS IMAGINE® Easytrace add-on. Easytrace helps reduce mouse clicks significantly while achieving similar or better digitizing results than classical methods (creating polygons, polylines, etc.).

153

Learning Objectives

1. To demonstrate a method of image interpretation through digitizing image features.
2. Creating polygons from a delineation approach using ERDAS IMAGINE's "Easytrace" option.
3. Creating polyline features to delineate boundaries using "Easytrace"

Data Required: **Y1326.tif**

This dataset represents a high-ground spatial resolution (30 cm [~1 foot]) and a 4-band multispectral dataset. The Y1326.tif dataset was obtained from the U.S. Fish and Wildlife Service's Blackwater National Wildlife Refuge, Maryland.

The refuge represents an area of national importance and has been referred to as the "Everglades of the North" by the Nature Conservancy. This refuge contains one-third of Maryland's tidal wetlands and maintains an incredible amount of plant and animal diversity. To find out more about the Blackwater National Wildlife Refuge and to download the data used in this exercise, visit the following website: https://www.fws.gov/refuge/Blackwater/. Components of this exercise adapted from ERDAS on-line documentation and the IMAGINE Easytrace User's Guide (ERDAS 2010b).

Blackwater National Wildlife Refuge Study Area (Figure 8.1).

FIGURE 8.1
Land cover type designations within the Y1326.tif image of the U.S. Fish and Wildlife Service's Blackwater National Wildlife Refuge, Maryland.

Image Digitizing and Interpretation 155

Polygon Creation

1. In an ERDAS IMAGINE file menu, select **File | Open | Raster Layer...** to open the **Select Layer to Add** dialog box. From the examples folder, select **Y1326.tif**.

   Notice that this file is in "**TIFF**" format, thus you must select "**TIFF**" as the file type when opening, or the file will not show up in your browser (Figure 8.2).

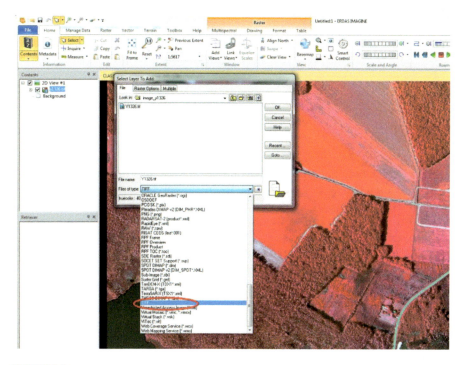

FIGURE 8.2
ERDAS IMAGINE import for a TIFF (*.tif) file format. (From ERDAS IMAGINE®/Hexagon Geospatial.)

2. In the file menu, select **File | New | 2D View #1 New Options | Vector Layer...** This will open the Create a New Vector Layer dialog box (Figure 8.3).

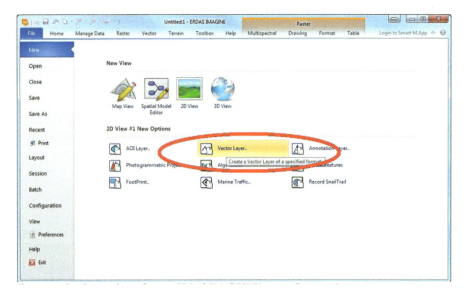

FIGURE 8.3
Load vector layer. (From ERDAS IMAGINE®/Hexagon Geospatial.)

3. In the Create a New Vector Layer dialog box, navigate to the directory where you would like to create your newly digitized data. The first feature that will be digitized will be a cultivated or agricultural field. Type the name **crop_trace** in the file name box and then select "**OK**" (Figure 8.4).

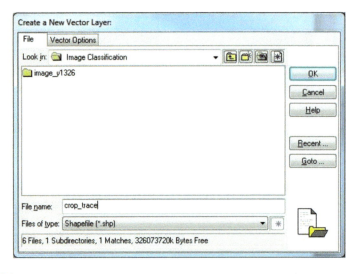

FIGURE 8.4
Create a new vector layer dialog window. (From ERDAS IMAGINE®/Hexagon Geospatial.)

Image Digitizing and Interpretation 157

4. In the "**New Shapefile Layer Option**" dialog box that appears, change the shapefile type to **Polygon Shape** and then select "**OK**" (Figure 8.5).

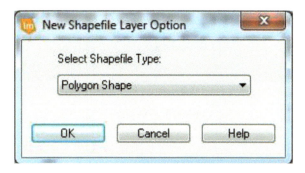

FIGURE 8.5
New shapefile layer option. (From ERDAS IMAGINE®/Hexagon Geospatial.)

5. Under the Vector menu tab group, click on the **EasyTrace** tool (**Drawing** tab | **Insert Geometry** grouping | **EasyTrace**) to access the tool (Figure 8.6).

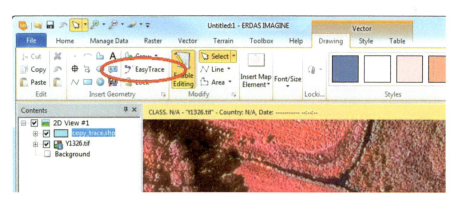

FIGURE 8.6
Load EasyTrace tool. (From ERDAS IMAGINE®/Hexagon Geospatial.)

6. In the Easytrace dialog window that opens, place a check next to the **Easy Tracing** box. This will make the "*Feature Type*" and "*Tracing Mode*" options available for use the following steps (Figure 8.7).

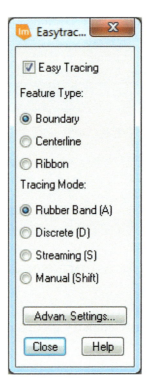

FIGURE 8.7
EasyTrace tool options. (From ERDAS IMAGINE®/Hexagon Geospatial.)

7. Under the Feature Type extraction options, if not all ready checked, select **Boundary**. This option sets the ability to more correctly digitize the border of land cover types within the imagery.
8. Under the Tracing Mode options, select **Rubber Band**. This option provides the ability to add vertices along the feature being digitized as the cursor is moved as the mouse button is clicked.
9. Click on the **Advan. Settings** button. The **Advanced Easytrace Settings** dialog that appears contains three weighting factors; Smoothness, Straightness and Image Feature. Increasing or decreasing these values will depend on how smooth or straight the image feature being digitized is. As well, if you prefer the software may be able to put more or less emphasis on helping to guide your digitization based on the actual underlying image feature that you are digitizing. For this example, change the Image Feature to "**93**." The increased Image Feature value, in comparison to the lower Smoothness and Straightness values, will allow the software to put greater emphasis on guiding the digitization of the agricultural field (Figure 8.8).

Image Digitizing and Interpretation 159

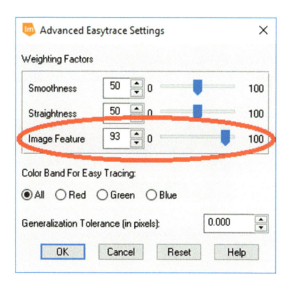

FIGURE 8.8
Advanced EasyTrace settings. (From ERDAS IMAGINE®/Hexagon Geospatial.)

10. Set the Color Band for Easy Tracing option to "**All**." This includes the visible bands and the IR band, which mimics the False Color Composite (FCC) configuration (4, 3, 2) that the image is being displayed in.
11. Depending on what classes you may be interested in digitizing (area land features, roads, etc.) the weighting factors can be adjusted to more correctly emphasize those features. This may involve some trial and error. For now, click on "**OK**" to close the dialog and accept the weighting factors. The Easytrace dialog window should remain open.
12. Locate an area in the image with agricultural cover and zoom into that area so that the entire boundary containing the area is clearly visible in the viewer.
13. Click on the **Polygon** icon from the Insert Geometry grouping in the Drawing Tabbed Ribbon section of the ERDAS IMAGINE file menu to digitize the agricultural field (polygon) in the image (Figure 8.9).

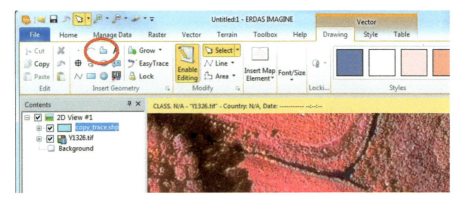

FIGURE 8.9
Polygon tool. (From ERDAS IMAGINE®/Hexagon Geospatial.)

14. Next, start tracing your agriculture feature by single-clicking along the border of your agricultural field. A digitized line will automatically be generated following your cursor. Add single-clicks to insert vertices where the generated line becomes less correct in following your cursor, and continue moving your cursor along the boundary to complete your digitization.
15. When you have completely digitized the entire surrounding boundary of your agricultural field, **double-click** the left mouse button to confirm the polygon. Open the attribute table and view the area and perimeter fields.
16. Repeat this procedure to delineate ALL agricultural uses within the image. Using the Rubber Band tracing mode with the weighting factor set to highly emphasize image features may still incorrectly follow your desired boundary line. Switching to the Manual tracing mode may offer greater control. While digitizing, you can change modes using the following **Hot Keys**:

    A—Rubber band

    S—Streaming

    D—Discrete

    Shift—Manual

    Z or **Backspace**—Undo

    C—To close polygon

Please refer to the Easytrace tool help in ERDAS IMAGINE, accessed from the bottom of the Easytrace tool dialog window, for more information on operation of each of these feature tracing modes.

Image Digitizing and Interpretation 161

Polyline Creation

1. In the Viewer, select **File | New | 2D View | Vector Layer...** This will open the Create a New Vector Layer dialog. In the Create a New Vector Layer dialog box, navigate to the directory where you created your agriculture field digitized data. The next feature that will be digitized will be a road feature. Type the name **road_trace** in the file name box. In the "**New Shapefile Layer Option**" dialog box that appears, change the shapefile type to Arc Shape and then select "**OK**."

2. If Easytrace is not already open or you have closed it, click on the **Easytrace** Tool icon (**Drawing | Insert Geometry | EasyTrace**).

3. If you had to re-open Easytrace in the previous step, check on the **Easy Tracing** box to enable Feature Type and Tracing Mode options.

4. Under the Feature Type extraction options, change the enabled selection to **Centerline**. This option sets the ability to more correctly digitize the center line of straight or curved linear features within the imagery.

5. Under the Tracing Mode options, select **Rubber Band**. This option provides the ability to add vertices along the feature being digitized as the cursor is moved as the mouse button is clicked. In the **Advanced Easytrace Settings** dialog that appears increase the Smoothness weighting factor to "**90**," increase the Straightness weighting factor to "**70**," and decrease the Image Feature weighting factor to "**34**." By increasing the Smoothness and Straightness values and decreasing the Image Feature values, the software may be able to put more emphasis on helping to guide your digitization along the road features that are relatively linear and smooth based on the actual underlying image feature that you are digitizing (Figure 8.10).

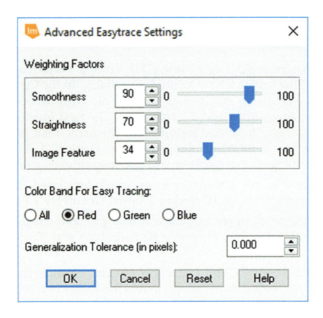

FIGURE 8.10
Advanced EasyTrace Settings for adding more smoothness and straightness and decreasing the Image Feature. (From ERDAS IMAGINE®/Hexagon Geospatial.)

6. Set the Color Band for Easy Tracing option to "**Red**." This will allow the software to use the red image band to help distinguish the contrast between the intended road features and non-road features.
7. Depending on what classes you may be interested in digitizing (area land features, roads, etc.) the weighting factors can be adjusted to more correctly emphasize those features. This may involve some trial and error. For now, click on "**OK**" to close the dialog and accept the weighting factors. The Easytrace dialog window should remain open.
8. Locate an area in the image with a road feature and zoom into that area so that the feature is clearly visible in the viewer.
9. Click on the **Polyline** icon from the Insert Geometry grouping in the Drawing Tabbed Ribbon section of the ERDAS IMAGINE file menu to digitize a road feature (line) in the image.
10. Create a template of the road's cross-section. In most urban and semi-urban areas, paved (and some unpaved roads) will maintain a consistent width (i.e., single lane, double lane, etc.). The consistent width of the road, as well as the texture within the feature, may be used by the software to assist with the digitization. To create a template of the road's cross-section, single-click on the edge of one side of the road, and then single-click again on the edge on opposite side of the road

Image Digitizing and Interpretation 163

(**careful not to double-click!**). This will generate a measurement template of the road's width or cross-section that is perpendicular to the road feature to be digitized.
11. Next, start tracing your road feature by single-clicking at the start of your road and moving your mouse cursor slowly along the center line of your road towards the end. A digitized line will automatically be generated following your cursor. Add single-clicks where the road becomes less smooth or straight (i.e., to move around a curve in the road).
12. When you are satisfied with the results, click the left mouse button to confirm the segment.
13. Double-click to end the digitization of the line. Open the attribute table and view the length field you have created.

Repeat this procedure to delineate ALL transportation routes within the image (delineation of the large road feature shown in the following image) (Figure 8.11).

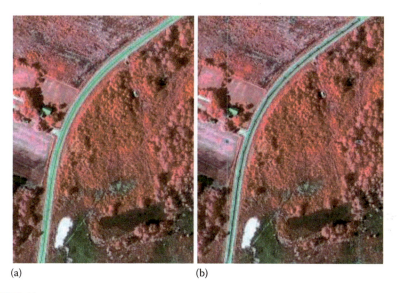

(a) (b)

FIGURE 8.11
Digitization of road features. Undigitized road feature (a) and digitized road feature (b). (From ERDAS IMAGINE®/Hexagon Geospatial.)

164 Image Processing and Data Analysis with ERDAS IMAGINE®

NOTE: The Hot Key "**Z**" (undo) or the backspace key on your computer keyboard may be used to correct any misplaced vertices generated while digitizing. You can also use the Hot Key "**Z**" or backspace key repeatedly to remove previous sets of vertices. If the feature has been completely digitized, the feature may be edited by selecting the desired feature (line or area) in the 2D View window and selecting either "**Line**" or "**Area**" from the Modify grouping in the Drawing Tabbed Ribbon section of the ERDAS IMAGINE file menu, and the selecting "**Reshape**." Individual vertices may be moved to a new position by clicking and dragging on a vertex. To complete the edits, simply click outside of the selected feature (Figure 8.12).

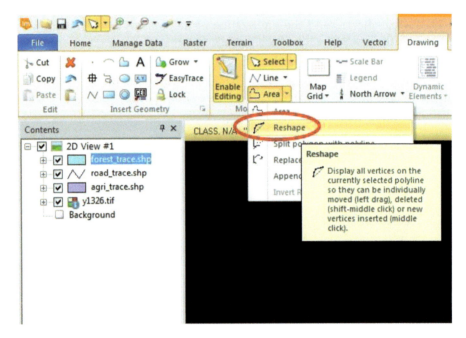

FIGURE 8.12
Reshape tool. (From ERDAS IMAGINE®/Hexagon Geospatial.)

Manual Drawing

All of the earlier procedures can be done WITHOUT using Easytrace. Try to create a few polygons/Polylines WITHOUT EasyTrace.

 1. Close Easytrace and in the file menu, select **File | New | 2D View #1 New Options | Vector Layer**... This will open the Create a New Vector Layer dialog box. Type the name **forest_trace** in the file name box and then select "OK."

Image Digitizing and Interpretation 165

2. In the "**New Shapefile Layer Option**" dialog box that appears, change the shapefile type to **Polygon Shape** and then select "**OK**."

3. Zoom into an area in the image with forest cover.

4. Click on the **Polygon** icon from the Insert Geometry grouping in the Drawing Tabbed Ribbon section of the ERDAS IMAGINE file menu to digitize a forest feature (polygon) in the image.

5. Begin single-clicking along the boundary of the forest taking care to set vertices in the appropriate places. When you are satisfied with the results, double-click to end the digitization of the polygon. Open the attribute table and view the area and perimeter fields you have created.

6. Repeat this procedure to delineate ALL forests within the image.

7. Now you may use either Easytrace or manual approach to continue delineating these other four land-use types:

   - Water (create shapefile "water.shp")
   - Wetland (create shapefile "wetland.shp")
   - Shrub/scrub (create shapefile "shrub_scrub.shp")
   - Bare Ground (create shapefile "bare_ground.shp")

Changing the Display Properties of the Digitizing Results

1. Select the vector layer file in your Content list that you would like to symbolize (crop_trace.shp, road_trace.shp, forest_trace.shp, etc.). Next select your color of choice from the the **Styles** grouping in the Style Tabbed Ribbon section of the ERDAS IMAGINE file menu.

   NOTE: In earlier versions of ERDAS IMAGINE (such as versions 2010–2013) the Viewing Properties options were located under the vector symbology button within the Viewing Properties grouping, i.e., **Vector | Style tab | Viewing Properties**, as seen in the figure below (Figure 8.13).

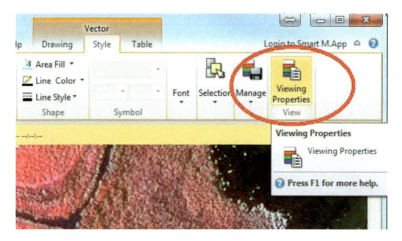

FIGURE 8.13
Viewing properties. (From ERDAS IMAGINE®/Hexagon Geospatial.)

2. Experiment with the different options in this window. For instance, change the polygon and road colors of each shapefile you created.

Review Questions

1. Explain how the process of digitization works as a method of providing data input.
2. Often the database created from manual digitization is stored as vector information. Explain the attributes of vector information.
3. Modern GIS and image processing software packages now allow for a "heads-up" digitization method. Explain this type of digitization method.
4. Modern GIS and image processing software also allows for a semi-automated feature extraction technique. What would be an advantage of this technique over other common methods?
5. Would the digitization process seem more effective as a data input method for an area with small, unevenly distributed, and non-contiguous features, or an area over the same size, but with larger, homogenous, and connect features?

9

Unsupervised Classification

Overview

The pixel values contained within raw digital, raster imagery serves as a rich data source for determining, quantifying, and mapping existing land use and land cover areas on the ground. As well as assessing, monitoring, or modeling changes in these areas over time. The operation of grouping or sorting image pixels, typically based on the pixel spectral values into categories (or themes), is known as image classification. In the image classification process, each *classified* category represents a unique land cover class that is representative of identified features on the ground. The resultant image classification map, or *categorized data*, is known as a thematic map. Two traditional image classification methods are known as unsupervised classification and supervised classification. The exercise associated with this chapter's application focuses on developing an unsupervised classification.

Unsupervised Classification: The unsupervised method of classification typically requires little or no input from the image analyst in developing the output land use/land cover classification. In this method, the classification system, or classifier, uses statistical means and covariance matrices to iteratively assign each pixel to a designated output class based on how spectrally separate each group of clustered pixels are. The analyst is then required to assign labels to the output categorical groups to complete the classification (see Table 9.1). The Iterative Self-Organizing Data Analysis Technique (ISODATA), as well as other standard clustering algorithms, represent common techniques for unsupervised classifications (Khorram et al. 2012b, Khorram et al. 2016b).

The ISODATA utilizes the mean and standard deviation in a number of bands in n-dimensional space (Tou and Gonzalez 1977, Sabins 1987) to iteratively group similar pixels. The user will guide the operation by selecting initial parameters, which include the maximum number of clusters, maximum percentage of pixels that can remain unchanged between iterations, maximum number of iterations, minimum percentage of pixels assigned to each cluster, maximum standard deviation, and minimum distance between clusters (Jensen 2005, Hester et al. 2008).

167

168 *Image Processing and Data Analysis with ERDAS IMAGINE®*

TABLE 9.1

Land Cover Classification Class Descriptions for the Unsupervised Classification of Paris, France Region

Land Cover Classification	Classification Description
Urban/ Development	Contained industrial, commercial, and private building, and the associated parking lots, a mixture of residential buildings, streets, lawns, trees, isolated residential structures or buildings surrounded by larger vegetative land covers, major road and rail networks outside of the predominant residential areas, large homogeneous impervious surfaces, including parking structures, large office buildings, and residential housing developments containing clusters of cul-de-sacs.
Agriculture/ Herbaceous Vegetation	Contained agricultural field, grass fields, and urban grasses. Characteristics of this class include large agricultural field, mowed/maintained lawns, fields, and vegetated road medians.
Evergreen Vegetation	Contained large homogeneous vegetative land covers of trees or shrubs that keep their leaves throughout the winter (mostly coniferous species).
Deciduous Vegetation	Contained large homogeneous vegetative land covers of trees or shrubs that lost their leaves during the winter (mostly hardwood species).
Barren Land	Contained areas or fields with little or no vegetation. Characteristics of this class include fallow agricultural fields, bare sediment or soil areas, and areas cleared of vegetation for construction.
Water	Contained lakes and small ponds and all other natural and artificial surface waters outside of the USGS National Hydrography Dataset shapefile.
Unclassified	Contained areas of unclassified or misclassified data.

Source: Hester D.B. et al., *Photogramm. Eng. Remote Sensing*, 74, 463–471, 2008.

Similar to the ISODATA approach, other clustering algorithms, such as the k-means clustering approach, also work in an iterative process. The initial stage includes the buildup of several similar pixel clusters within a range determined by the analyst. Each resulting cluster is composed of pixel groups consisting of similar spectral values, which likewise occupies a common spectral space (Jain 1989, Celik 2009) that consists of a well-defined mean vector for each class. This operation is followed by a minimum-distance-to-means classification algorithm that determines the final pixel cluster groups. In the case of the k-means clustering algorithm, the data space is partitioned into Voronoi cells. The Voronoi cells represent regions that each contain sets of points, or seeds, which correspond to the cluster means that are closest in multidimensional Euclidean space. In the k-means operation, pixels are iteratively classified into a predetermined number of clusters with no deletion, splitting, or merging of the clusters between each iteration.

I. Unsupervised Classification Application

Learning Objectives

1. To gain an understanding of the general procedures for image classification using the unsupervised image analysis process to classify land use and land cover.
2. To gain an understanding of the general procedures for image classification using the unsupervised image analysis process to classify land use and land cover.

Data Required: This exercise will demonstrate an unsupervised classification approach on a Landsat 8 Operational Land Imager/Thermal Infrared Sensor (OLI/TIRS) image of Paris France, downloaded from the United States Geological Survey (USGS) EarthExplorer (http://earthexplorer.usgs.gov/) website as previously described in Chapter 1.

Obtaining the Required Data (Review)

From the USGS EarthExplorer website, this exercise will download the Landsat 8 image: **LC08_L1TP_199026_20170527_20170615_01_T1**, Acquisition Date: 2017/05/27 for the Paris, France area.

1. In the Address/Place field, type in the word *"Paris"* and click on the **"Show"** button below the box. Next, click on the Paris, France link under the Address/Place heading to show the location on the map and add coordinates to the Area of Interest Control.
2. In the **"Search from"** boxes in the Search Criteria section enter the following dates: **05/01/2017** to **08/31/2017**. Next select **"Data Sets"** below Date Range options.
3. Select **"Landsat Archive | L8 OLI/TIRS C1 Level-1"** on the Data Sets tab.
4. Next select **"Results"** at the bottom of Data Sets tab. Now inspect the available imagery.

   As previously noted, the majority of available L8 OLI/TIRS products (representing the Landsat 8 Operational Land Imager and Thermal Infrared Sensor) in this image are obscured by cloud cover. This imagery would make land cover image analysis very difficult. It is best to try to locate imagery within the target area with as minimal cloud cover as possible. However, if particular areas that you may be interested in within the image are not obscured by clouds, then this image may still be usable (Figure 9.1).

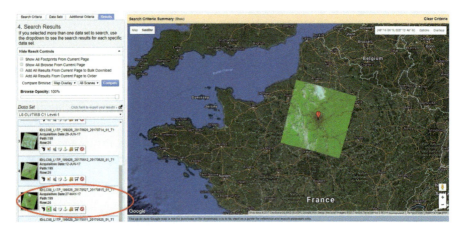

FIGURE 9.1
EarthExplorer data search for the Landsat 8 OLI/TIRS image of the Paris, France area (File Name: LC08_L1TP_199026_20170527_20170615_01_T1, Acquisition Date: 2017/05/27).

Next examine the **LC08_L1TP_199026_20170527_20170615_01_T1** image.

5. Click on the **Show Metadata and Browse** icon for the LC08_L1T P_199026_20170527_20170615_01_T1 image.

Here you can visually inspect a view of the image in the metadata table that opens. Also, by scrolling down the metadata table, note that the **Image Quality = 9** (or best quality), and the **Land Cloud Cover** (the percentage of land image data that is obscured by clouds) is approximately 19% (Figure 9.2).

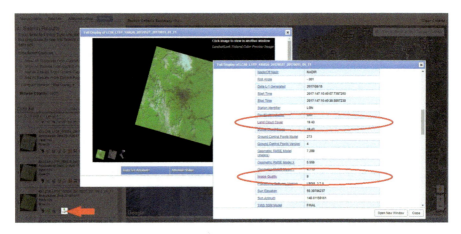

FIGURE 9.2
Land cloud cover inspection.

Unsupervised Classification 171

Theoretically, a Land Cloud Cover percentage of 19% means that 81% of the image should be unobscured by clouds. It is recommended to focus the classification on an area within the image that is unobscured by clouds.

6. To download the image, select the **Download Options** icon ⬇ to begin downloading the Landsat 8 product (LC08_L1TP_199026_2017 0527_20170615_01_T1).

Continue to follow the directions discussed previously in **Chapter 1** to unzip the image (7. "Unzipping the L8 OLI/TIRS Data") and create a multi-band image (10. "Creating a Layer Stacked Image (Landsat 8 data)") in ERDAS IMAGINE®.

NOTE: Remember it is not necessary to add the following files to the Layer Stack image for this exercise, as these files represent either panchromatic data (Band 8) or thermal infrared data (Bands 10 and 11), as well these bands are captured at different spatial resolutions:

LLC08_L1TP_199026_20170527_20170615_01_T1_B8.TIF (Panchromatic)

LC08_L1TP_199026_20170527_20170615_01_T1_B10.TIF (Thermal Infrared 1)

LC08_L1TP_199026_20170527_20170615_01_T1_B11.TIF (Thermal Infrared 2)

The new multi-band image will consist of bands 1–7 and band 9 (the original band 9 will be called band 8 in the new multi-band image). Remember that Landsat 8's band 9 will be represented in the multi-band image as band 8, since Landsat 8's band 8 was not included in the layer stack operation.

7. Name the output file: **LC08_L1TP_199026_20170527_20170615 _01_T1_ALB** and change the output file type to "Files of type: **IMAGINE Image (*.img)**" from the drop-down list. Note the "**ALB**" suffix represents a designation that the newly created file will now be comprised of "All Bands" (Bands 1–7, and 9). Also, be sure to check "**Ignore Zero in Stats.**" Finally, click on "**OK**" to begin the layer stack conversion and dismiss the Process List once the Layer Stack is complete.

The Layer Stack band selections should resemble Figure 9.3.

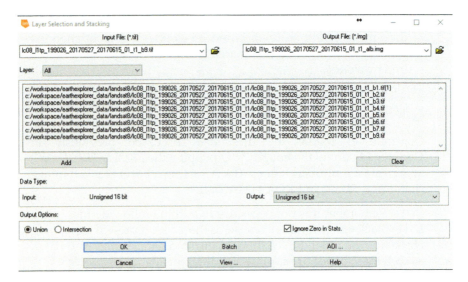

FIGURE 9.3
Layer stack operation of the Landsat 8 OLR/TIRS bands 1–7 and 9. (From ERDAS IMAGINE®/Hexagon Geospatial.)

Creating an Image Subset of the Cloud-Free Areas

To create an area useful for image classification, it will be necessary to create an image subset of the cloud-free regions within the newly created multi-band image.

NOTE: In Chapter 1, the previously downloaded Sentinel-2 MSI, cloud-free image for the Paris, France region (File Name: **L1C_T31UDQ_ A010058_20170526T105518**, Acquisition Date: 2017/05/26) closely matches the date of the new Landsat 8 image. The area unobscured by clouds, which will be selected, is cloud-free in both the Sentinel-2 MSI and Landsat 8 images (Figure 9.4).

Unsupervised Classification 173

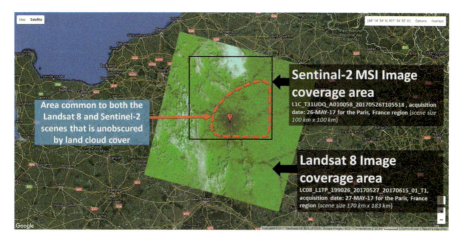

FIGURE 9.4
Comparison of the cloud free areas common in both the Landsat 8 OLI/TIRS imagery (Acquisition Date: 2017/05/27) and the Sentinel-2 MSI imagery (Acquisition Date: 2017/05/26) of the Paris France region.

1. Open the new multi-band image, LC08_L1TP_199026_20170527_20170615_01_T1_ALB.img, as a False Color Composite (FCC) display as previously discussed in Chapter 2 (Band Combinations) using the following band combination: 5 (red), 4 (green), 3 (blue).

2. Using the zoom tool from the main menu ribbon, zoom into the center of the image and inspect the area for cloud cover. This will ensure the selection of an area that is unobscured by clouds.

3. Next, select **Inquire** icon from the main menu ribbon and then **Inquire Box**. Click-drag the white inquire box in the 2D View window to encompass an area free from clouds. Then click on "Apply." Leave the inquire box open (Figure 9.5).

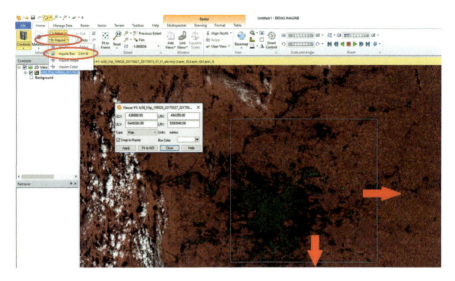

FIGURE 9.5
Area of image subset. (From ERDAS IMAGINE®/Hexagon Geospatial.)

4. Next, from the main menu bar, select the **Raster** tab. From the menu ribbon, select **Subset & Chip | Create Subset Image**. In the Subset dialog window that opens, the Input file should be populated with the multi-image Landsat scene currently displayed in the 2D View window. In the Subset Definition section, click on the "**From Inquire Box**" button. This should automatically load the coordinates from the inquire box. Name the Output File: **lc08_l1tp_199026_20170527 _20170615_01_t1_subset.img**. Finally, select "**Ignore Zero in Stats**" and "**OK**" to begin the subset operation (Figure 9.6).

Unsupervised Classification 175

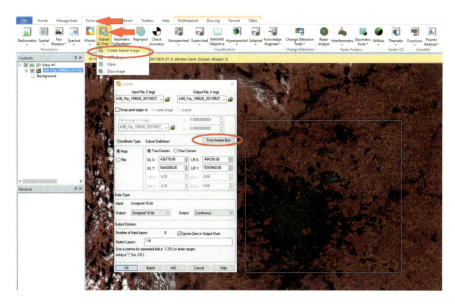

FIGURE 9.6
Create image subset options. (From ERDAS IMAGINE®/Hexagon Geospatial.)

5. Clear the 2D View display window by right-clicking anywhere in the 2D View window and selecting "**Clear View**" from the pop-up list that appears.
6. Next, open the new subset image as a False Color Composite (FCC).

File | Open | Raster Layer. In the "**Select Layer to Add:**" window that opens, navigate to where you stored the new subset image. Next, click on the Raster Options tab and in the "Layers to Colors" section enter the following band combination: 5 (red), 4 (green), 3 (blue). Check the "**Fit to Frame**" option. Then click on "**OK**" (Figure 9.7).

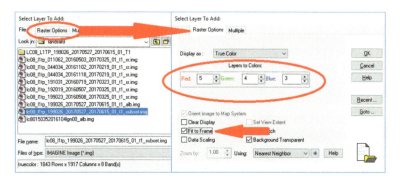

FIGURE 9.7
Raster layer display options. (From ERDAS IMAGINE®/Hexagon Geospatial.)

The result is a multi-band image that can use to begin the classification process (Figure 9.8).

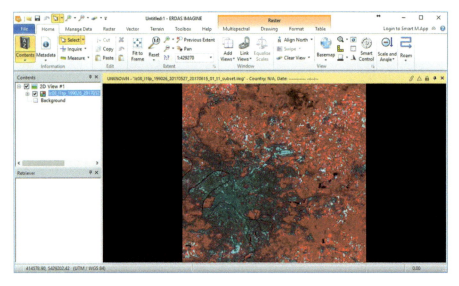

FIGURE 9.8
Multi-band image result. (From ERDAS IMAGINE®/Hexagon Geospatial.)

Initiating the Unsupervised Classification

When opening the image in ERDAS IMAGINE, check the "**Raster Options**" tab to ensure you apply the standard FCC band combination (5, 4, 3) to the displayed image. Use the **Metadata** icon from the Home tab to obtain information about the image file. Take note of the appearance of specific ground features displayed with the data using the standard FCC band combination (such as the color of vegetation within the image), and whether the composite appears to enhance any specific feature.

For the unsupervised classification, you should try to identify the following five classes ("unclassified" will be included by default as a sixth class), along with the specified color and definition scheme.

Land Cover Category and Color to apply:

Class 1: Water—Blue

Class 2: Urban/Developed Land—Cyan

Class 3: Agriculture/Herbaceous Vegetation—Yellow

Class 4: Forested Vegetation—Dark Green

Unsupervised Classification 177

Class 5: Barren Land—Orange

Class 6: Unclassified—Black

Unsupervised Classification Application Approach

1. Select **Raster | Unsupervised** within the Classification category grouping **| Unsupervised Classification**.

2. In the "**Unsupervised Classification**" dialog window that opens, select the "**Help**" button to become more familiar with the available options this tool offers.

3. In the "**Unsupervised Classification**" dialog, select the following options:

   - **Input Raster File:** This field should be populated with the subset image that is now in the 2D View window. Alternatively, you may browse and select the image file you downloaded

   - **Output Cluster Layer:** Name the output file "*paris_may2017_unsupervised.img*" and save it to a folder that you can easily find (preferably where initial images have been saved). The default option will save the output file as an "**.img*" file, which is native format of ERDAS IMAGINE.

   - **Output Signature Set:** Leave this option unchecked.

   - **Number of Classes:** This will be the number of unsupervised output classes generated. For this demonstration, choose 200.

     NOTE: Ordinarily, when exploring potential classification categories, it is a good idea to start with a high number of automatically generated, unsupervised output classes for each class category that you are interested in achieving in the final thematic map. For example, if after inspecting the raw imagery, the intention is to classify the raw imagery into a minimum of six land cover class categories, an appropriate strategy would be to estimate roughly **200 or more** unsupervised output classes "**FOR EACH**" of the intended land cover classes you hope to achieve (e.g., **six final land cover classes × 200 unsupervised output classes each = 1,200 total output classes**). This large number of total output classes provides a greater *chance* of capturing the majority of the variability within the final six land cover classes in comparison to selecting a smaller number of output classes. Additionally, it is a good idea to use the spectral profile tool to examine representative classes throughout the entire raw image to get an idea of how similar or variable the spectral values may be for each class (Figure 9.9).

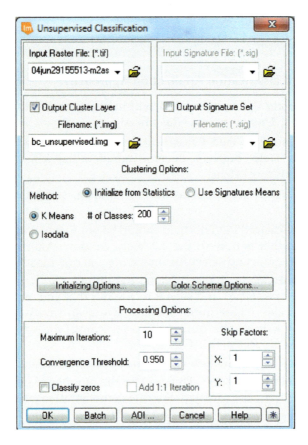

FIGURE 9.9
Unsupervised classification options. (From ERDAS IMAGINE®/Hexagon Geospatial.)

Unsupervised Classification 179

Color Scheme Options…

ERDAS IMAGINE provides a default Color Scheme Option that allows the output unsupervised classification to resemble the original input data ("Approximate True Color" option shown in Figure 9.10).

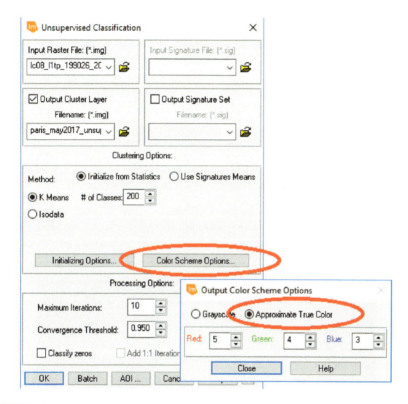

FIGURE 9.10
Unsupervised classification Color Scheme Options. (From ERDAS IMAGINE®/Hexagon Geospatial.)

Since the original input image was displayed as a FCC, the "**Approximate True Color**" option applied the unsupervised classification as an approximate of the false color reflectance values of the input image (see the following image). The can be of certain benefit to an image analyst, particularly if unfamiliar with the area being classified. However, this may also bias the analyst by essentially *suggesting* what the appropriate class designation should be (Figure 9.11).

180　　　*Image Processing and Data Analysis with ERDAS IMAGINE®*

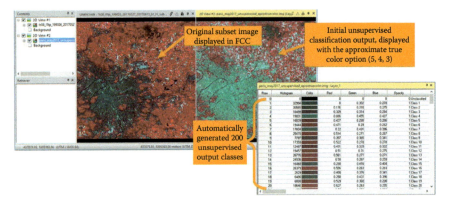

FIGURE 9.11
Original subset image (on left) and initial unsupervised classification result with 200 classes generated (on right). (From ERDAS IMAGINE®/Hexagon Geospatial.)

To conduct the classification as objectively as possible, this exercise will select a grayscale option for the output classification.

- Selected the "**Color Scheme Options…**" button and choose the "**Grayscale**" output option instead of the default "**Approximate True Color**" option as shown in the following image. Then click on "**close**" in the Output Color Scheme Options dialog box (Figure 9.12).

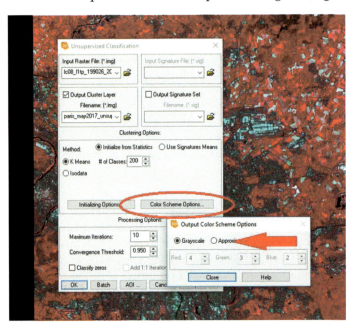

FIGURE 9.12
Grayscale color scheme options. (From ERDAS IMAGINE®/Hexagon Geospatial.)

Unsupervised Classification 181

Complete the Unsupervised Classification properties dialog box by choosing the following options for "**Maximum Iterations**" and "**Convergence Threshold**."

- **Maximum Iterations:** 10
- **Convergence Threshold:** 0.95
- Click **OK**.

A Process List window should appear and begin the procedure. Once processing is complete, **close** the process list window.

4. Once you have completed the classification, open another **2D Viewer** and display the newly created unsupervised image in this viewer.
5. In the viewer contents list (table of contents on the left), right-click on the newly created unsupervised image's file name and select "**Display Attribute Table**" (Figure 9.13).

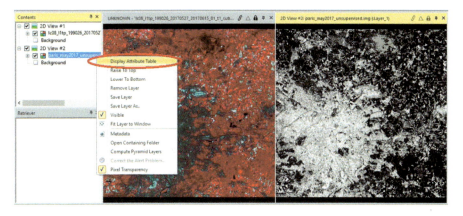

FIGURE 9.13
Original subset image (on left) and grayscale unsupervised classification result (on right). (From ERDAS IMAGINE®/Hexagon Geospatial.)

An attribute, similar to the following figure, should open at the bottom of the screen. You may reposition this table by dragging the table's header portion of the frame into the main graphic user interface (GUI), or other Window's space on the desktop, so that you can expand it further (Figure 9.14).

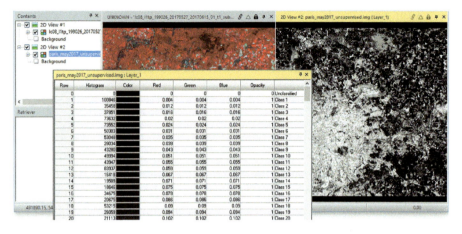

FIGURE 9.14
Attribute table of grayscale unsupervised classification result (on right). (From ERDAS IMAGINE®/Hexagon Geospatial.)

From this point, you would begin **"working"** through each output class generated by the unsupervised classification algorithm by turning the color of the first output class (class 1) to something you can easily identify (such as yellow or red) and then comparing the highlighted color to the **"reference image**." This color is considered an **Alarm Color**. The Alarm Color is a temporary color on the unsupervised classification that allows you to further investigate the class from the original reference image.

Creating a Reference Image Subset

Ideally, a reference image should be an independent image source, covering the same location, captured at the same time, and of a higher resolution than the classification data. This ensures that the classification is not biased by spatial or temporal differences, as well as serving to not to statistically compromise the *interpretation*. Fortunately, the previously downloaded Sentinel-2 MSI, cloud free, image for the Paris, France region (File Name: L1C_T31UDQ_A010058_20170526T105518, Acquisition Date: 2017/05/26) closely matches the date for the Landsat 8 image (Chapter 1 "Finding and Downloading Data in EarthExplorer"). Also, the Sentinel-2 image is of a higher spatial resolution. Remember in Chapter 1, a multi-band image was also created (**T31UDQ_20170526T105031_B02-4_8**) from the original

Unsupervised Classification 183

Sentinel-2 MSI data. The multi-band image will be used to create the reference image subset by repeating the aforementioned procedures (see *Creating an Image Subset of the Cloud-Free Areas*) to subset the corresponding area on the Sentinel-2 MSI data that represents the area unobscured by clouds.

NOTE: After adding Sentinel-2 MSI data (t31udq_20170526t105518_b02-4_8) to the display, the Landsat 8 data subset (lc08_l1tp_199026_20170527_20170 615_01_t1_subset.img) can be opened in the same 2D View window to use for duplicating the Inquire Box dimensions. Name the new Sentinel-2 subset image: t31udq_20170526t105031_b02-4_8subset.img (Figure 9.15).

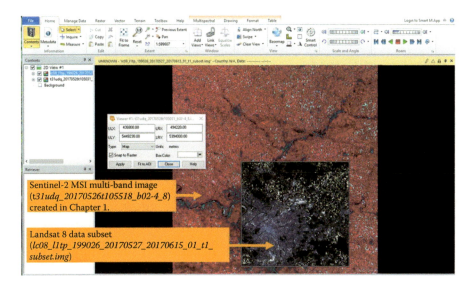

FIGURE 9.15
Comparison of the extent of Landsat 8 subset image to the Sentinel-2 image. The Landsat subset image's extent will be used to subset the Sentinel-2 image. (From ERDAS IMAGINE®/Hexagon Geospatial.)

6. Clear the display of the multi-band Sentinal-2 data and load the new Sentinel-2 subset image (t31udq_20170526t105031_b02-4_8subset. img) in the same 2D View window as the Landsat 8 subset (lc08_l1tp_ 199026_20170527_20170615_01_t1_subset.img). Feel free to use the Swipe Tool to inspect the new image subset (Figure 9.16).

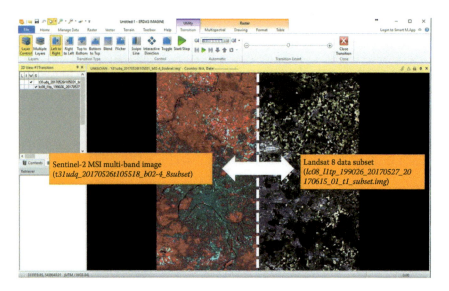

FIGURE 9.16
Swipe Tool comparison of the Landsat image subset to the Sentinel-2 image subset. (From ERDAS IMAGINE®/Hexagon Geospatial.)

Assigning Class Categories to the Unsupervised Classification

Remove both subset images from the 2D View window (right-click Clear View) and open Sentinel-2 MSI data as a True Color Composite (TCC) display (3-2-1, remember band 1 was not included in the original layer stack). Open a second 2D View window and add your raw Landsat 8 unsupervised classification. The Sentinel-2 subset image will serve as the reference image, and the raw Landsat 8 unsupervised classification data will serve as the input data for assigning individual class categories. In the viewer contents list (table of contents on the left), right-click on the file name of the input image (e.g., Landsat 8 unsupervised classification data) and select "**Display Attribute Table**" (Figure 9.17).

Unsupervised Classification 185

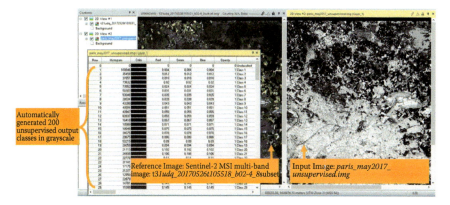

FIGURE 9.17
Attribute table of the grayscale Landsat unsupervised classification (right) used for assigning class categories to the unsupervised classification. (From ERDAS IMAGINE®/Hexagon Geospatial.)

7. Right-click in the "**Color**" of Class 1 to change the color from a dark gray to yellow.

    NOTE: "**Class 0**" should represent the background area of the image and should be labeled in the "Class_Names" column of the attribute table (the last column) as "**Unclassified**." Do not change this class (Figure 9.18).

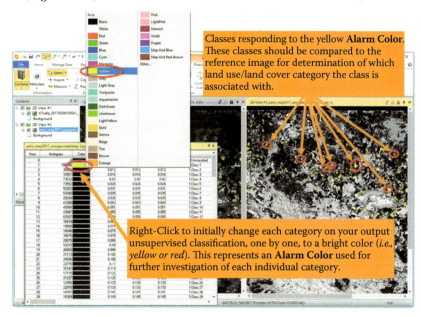

FIGURE 9.18
Assignment of class categories to the grayscale Landsat unsupervised classification (right) in the attribute table. (From ERDAS IMAGINE®/Hexagon Geospatial.)

8. Identify where this highlighted color (yellow) shows up on the gray-scale supervised classification image.
9. Compare the identified "class" to the same location in the reference image (Figure 9.19).

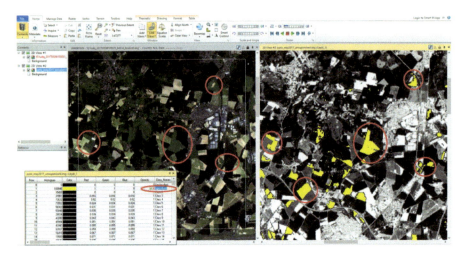

FIGURE 9.19
Comparison of identified classes (right) to the same location in the reference image (left) using yellow as an alarm color. (From ERDAS IMAGINE®/Hexagon Geospatial.)

NOTE: Linking the 2D Views might be helpful in confirming the location on the Sentinel-2 subset reference image (left 2D View).

10. Once you have identified the type of land cover that the highlighted class (yellow) is now displaying, type in the identified class category from the list of the five target class categories (Water, Urban/Developed Land, Agriculture/Herbaceous Vegetation, Forested Vegetation, and Barren Land) in the last column of the attribute table (Figure 9.20).

Unsupervised Classification 187

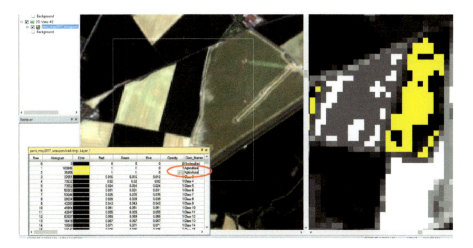

FIGURE 9.20
Assigning the appropriate land cover category in the attribute table. (From ERDAS IMAGINE®/Hexagon Geospatial.)

11. Then repeat this procedure "**FOR EACH**" of the remaining 199 output classes you requested the software to generate. For each spectral class (spectral cluster; row in the table), identify the cover-type represented in the new image (such as water, developed vegetation, etc.).

    NOTE: Only **200 output classes for the unsupervised classification** were specified. However, remember the classification results may be may be "**GREATLY IMPROVED**" by specifying a **higher** number of output classes for each land cover category desired the final thematic map. A good rule of thumb is to start with roughly 200 output classes for each intended category desired to achieve in the final unsupervised thematic map. Additionally, it is a good idea to use the spectral profile tool to examine representative classes throughout the entire raw image to get an idea of how similar or variable the spectral values may be for each class. However, for this example only 200 output classes will be used.

12. Once you have identified each appropriate land cover category, right-click on the "**Color**" in the attribute table and select a color change as suggested in the classification legend earlier (e.g., Land Cover Category and Color to apply: Water—Blue, Urban/Developed Land—Cyan, Agriculture/Herbaceous Vegetation—Yellow, Forested Vegetation—Dark Green, Barren Land—Orange).

    Remember to edit the class names and colors in the attribute table as you go through each class.

13. Save the edited unsupervised classification output image. Right-click on image file name in your Contents Legend and select "Save Layer."

   NOTE: Remember, classifications are **rarely perfect**! However, you should try to develop the **best** classification possible from the comparisons of the original data.

Compare

1. To visually compare the classification results, use the swipe tool.
2. Open the reference image in the same 2D Viewer as the unsupervised classification output image.
3. Rearrange the image order in the Contents Legend by dragging the unsupervised classification output image to the top of the list, so that the reference image is below the unsupervised classification output image (Figure 9.21).

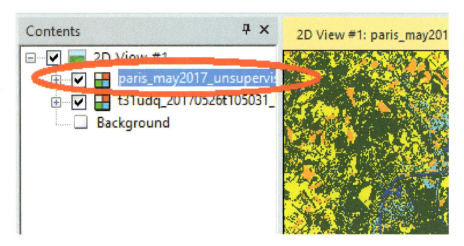

FIGURE 9.21
Rearrange the image order in the Contents Legend so that the newly completed unsupervised classification is on the top. (From ERDAS IMAGINE®/Hexagon Geospatial.)

4. Open the swipe tool (**Home** tab | **Swipe** within the View category grouping) to compare the land cover classification for areas of agreement and disagreement (Figures 9.22 and 9.23).

Unsupervised Classification 189

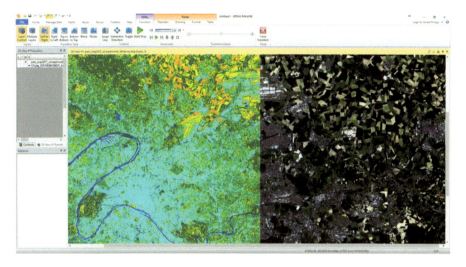

FIGURE 9.22
Swipe Tool comparison of the newly completed unsupervised classification and Sentinel-2 (reference image). (From ERDAS IMAGINE®/Hexagon Geospatial.)

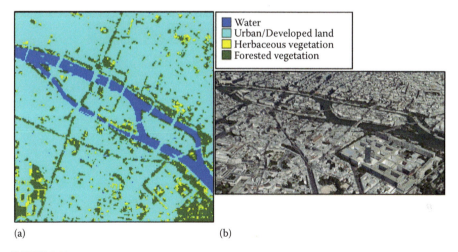

(a) (b)

FIGURE 9.23
Sentinel-2 subset image comparisons of the Isle of St Louis area near the city center of Paris, France (t31udq_20170526t105031_b02-4_8subset.img; Acquisition Date: 2017/05/26). (a) Unsupervised Classification result, using 300 input classes within the K-means clustering algorithm option in an effort to improve representation of four land cover classes; Water, Urban/Developed Land, Herbaceous Vegetation, and Forested Vegetation. (From ERDAS IMAGINE®/Hexagon Geospatial.) (b) Oblique/3D render of Google Earth image of the city center of Paris, France (image date 8/6/2016).

II. Unsupervised Classification in Esri ArcMap ArcGIS for Desktop

1. In ArcMap ArcGIS for Desktop, use the **Add Data** icon to add the multi-band image created in Chapter 1, Section III. *Displaying raster data and creating a multi-band image in Esri ArcMap ArcGIS for Desktop* (ArcMap_T31UDQ_20170526T105031_B02-4_8.tif) (Figure 9.24).

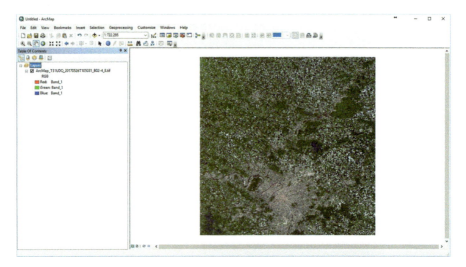

FIGURE 9.24
Esri ArcMap ArcG IS for Desktop application interface. (From ARCMap/ESRI.)

2. Use the **Zoom** tool to zoom into an area similar to the Sentinel-2 subset image created in the earlier section (t31udq_20170526t105031_b02-4_8subset.img) (Figure 9.25).

Unsupervised Classification 191

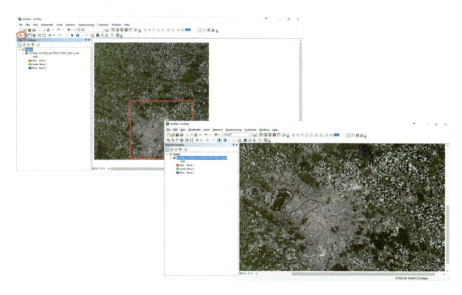

FIGURE 9.25
Image subset region (highlighted in red) created for the multi-band Sentinel-2 image (t31udq_20170526t105031_b02-4_8.img). (From ARCMap/ESRI.)

3. From the file menu at the top of the ArcMap interface, select **Window | Image Analysis**. In the Image Analysis window that opens, select the **Clip** icon from the Processing section (Figure 9.26).

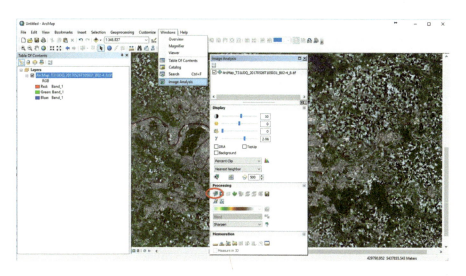

FIGURE 9.26
ArcMap Image analysis options. (From ARCMap/ESRI.)

4. Use the Data Export command (**Right-click** on the newly created "Clip_ArcMap_T31UDQ_20170526T105031_B02-4_8.tif" in the Table of Contents | **Data** | **Export Data**). This will open an "Export Raster Data" window. Navigate to the directory you wish to save the new file to and click on "Save." In the Output Raster window that pops up, click on "No" to prevent ArcMap from adding NoData pixels to the new file. Finally, add this new file to the map document by selecting "Yes" when presented with the option.
5. Rename the new file in the Table of Contents "**Clip_ArcMap_T31UDQ_20170526T105031_subset.tif**" and remove the other files layers within the Table of Contents.
6. Turn on the Spatial Analyst extension in ArcMap by placing a check beside the Spatial Analyst option in the Extensions menu (**Customize | Extensions… | Spatial Analyst**) (Figure 9.27).

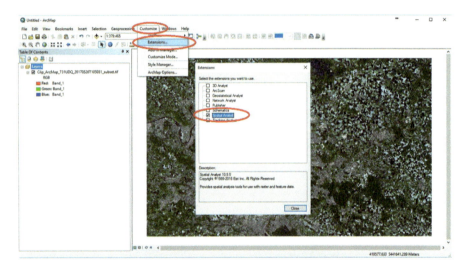

FIGURE 9.27
Spatial Analyst extension activation in ArcMap. (From ARCMap/ESRI.)

7. Next, open the ArcToolbox by clicking on the **ArcToolbox** icon in the file menu. In the ArcToolbox, navigate to Spatial Analyst Tools | Multivariate | ISODATA Cluster Unsupervised Classification. This will open the ISODATA Cluster Unsupervised Classification dialog window (Figure 9.28).

Unsupervised Classification 193

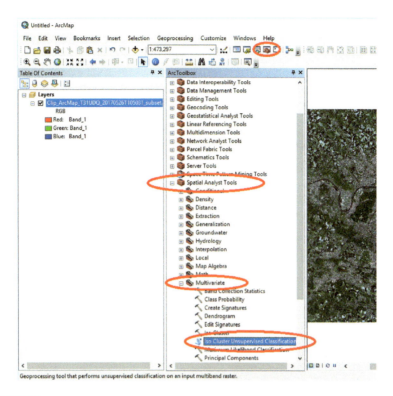

FIGURE 9.28
Isodata Cluster Unsupervised Classification tool selection in Arc ToolBox. (From ARCMap/ESRI.)

8. In the "ISODATA Cluster Unsupervised Classification" dialog, select the following options:

- **Input raster bands:** Drag the "**Clip_ArcMap_T31UDQ_20170526T105031_B02-4_8.tif**" layer from the Table of Contents into this field. Alternatively, you may browse and select the image file.

- **Number of classes:** This will be the number of unsupervised output classes generated. Esri recommends using a number equal to ten times the number of bands within the multi-band image. For this field designate **40**.

- **Output classified raster:** Leave this option the default, as Spatial Analyst will sometimes not operate correctly if the default location for processing is changed. The Export Data command can be used to rename and save output file as "*ArcMap_paris_may2017_unsupervised.tif*" so that this version indicates in the file name that it was completed in ArcMap.
- **Minimum class size:** Leave this option on the default of **20**.
- **Sample interval:** Leave this option on the default of **10**.
- **Output signature file:** Leave this option blank (Figure 9.29).

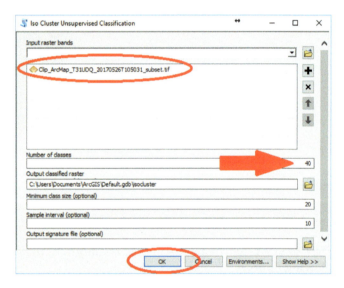

FIGURE 9.29
Isodata Cluster Unsupervised Classification dialog window in Arc ToolBox. (From ARCMap/ESRI.)

9. Once the unsupervised classification is complete, right-click on the classification file name (i.e., *isocluster*), select **Properties...** | **Symbology** tab within the Layer Properties window | **Classified** |, change the Classes to **32**, then click **OK** (Figure 9.30).

Unsupervised Classification 195

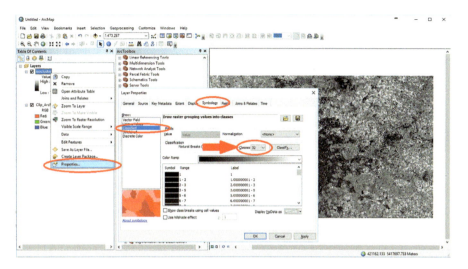

FIGURE 9.30
Output class number selection on the Symbology tab in Layer Properties dialog window. (From ARCMap/ESRI.)

> NOTE: Only **40 output classes for the unsupervised classification were specified**. No classification is ever 100% accurate. However, remember the classification results may be **"GREATLY IMPROVED"** by specifying a **higher** number of output classes for each desired land cover category that you would like to achieve in the thematic map. Additionally, it is a good idea to use the spectral profile tool to examine representative classes throughout the entire raw image to get an idea of how similar or variable the spectral values may be for each class.

10. Next, follow the unsupervised classification approach completed previously (see Unsupervised Classification Application Approach, Step 5). Begin **"working"** through each output class value generated by the unsupervised classification algorithm by turning the color of the first output class (value 1) to something you can easily identify (such as yellow or red). This color is considered an Alarm Color. The Alarm Color is a temporary color on the unsupervised classification that allows you to further investigate the class to correctly identify its true land cover.

To implement the first Alarm Color, right-click on the first value (value 1) under the isocluster layer in the Table of Contents. In the color pallet that opens, choose YELLOW. Identify the land cover in the classification that turns YELLOW. In the following image, the river turns YELLOW, so call this value water, and then select an appropriate color for water (such as BLUE).

NOTE: By unchecking the checkbox next to the unsupervised classification (e.g., isocluster) in the Table of Contents, you can toggle between the unsupervised classification layer and the subset layer, which may be useful as a reference.

For the unsupervised classification, you should try to identify the following five classes, along with the specified color and definition scheme:

Land Cover Category and Color to apply

Class 1: Water—Blue

Class 2: Urban/Developed Land—Cyan

Class 3: Agriculture/Herbaceous Vegetation—Yellow

Class 4: Forested Vegetation—Dark Green

Class 5: Barren Land—Orange

See Table 9.1 for a description of each land cover category (Figure 9.31).

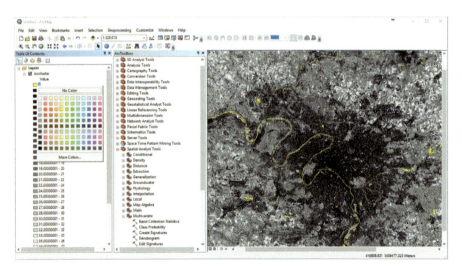

FIGURE 9.31
Evaluation of each output class value generated by the unsupervised classification algorithm. (From ARCMap/ESRI.)

11. Continue "**working**" through each output class value until you have assigned all 32 class values to one of the five land cover categories (Figure 9.32).

Unsupervised Classification 197

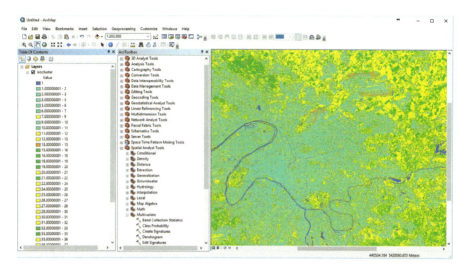

FIGURE 9.32
Completion of assigning all 32 class values to one of five land cover categories. (From ARCMap/ESRI.)

12. Evaluate the classification for any classes that may need further investigations or manual editing, such as misclassified classes.
13. Next, reclassify the 32 classes into the five land cover categories by selecting in Arc Toolbox: **Spatial Analyst Tools | Reclass | Reclassify**.

    In the Reclassify window that opens, drag the unsupervised classification layer into the **Input Raster field**. Alternatively, you can navigate to where this file is stored and loaded from there. For the **Reclass field** select "**Value**." Use the colors you selected for the 32 class values in the Table of Contents and change the values in the "New values" column to match each of the five land cover categories (e.g., 1 for Water [Blue], 2 Urban/Developed [Cyan], 3 for Agriculture/Herbaceous Vegetation [Yellow], 4 for Forested Vegetation [Dark Green], and 5 for Barren Land [Orange]) (Figure 9.33).

198 *Image Processing and Data Analysis with ERDAS IMAGINE®*

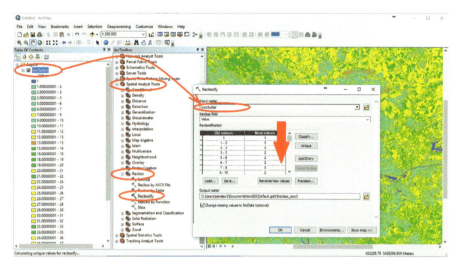

FIGURE 9.33
Reclassify tool used to reclassify the 32 unsupervised classification class values into the five land cover categories. (From ARCMap/ESRI.)

14. Once you have completed adding all the new class values, click on "**OK**" in the Reclassify window to convert the original 32 classes to the five land cover categories. Next, assign the appropriate colors to the five class values in the reclassified layer that has been loaded into the Table of Contents (Figure 9.34).

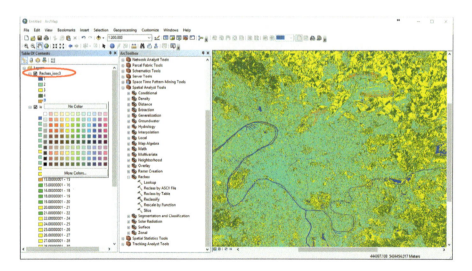

FIGURE 9.34
Selection of reclassification colors representative of the five land cover categories. (From ARCMap/ESRI.)

Unsupervised Classification 199

15. Finally, you may wish to use the Export Data command (right-click on *Reclass_isoc* in the Table of Contents | **Data** | **Export Data**) to rename and save output file as *"ArcMap_paris_may2017_unsupervised.tif"* so that this version can be identified in the file name as being completed in ArcMap. Also, you may wish to use the **File** | **Save As...** command in the file menu to save the ArcGIS ArcMap map document.

III. Unsupervised Classification in QGIS

1. Open QGIS Desktop (current version *QGIS Desktop 2.18.12 with GRASS 7.2.1*), from the file menu at the top of the QGIS interface, use the **Layer** | **Add Layer** | **Add Raster Layer...** option to navigate the directory to add the multi-band image created in Chapter 1, Section *IV. Displaying raster data and creating a multi-band image in QGIS* (QGIS_T31UDQ_20170526T105031_B02-4_8.tif) (Figure 9.35).

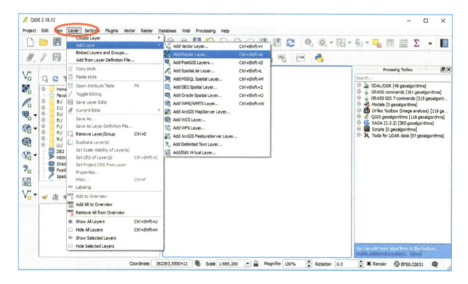

FIGURE 9.35
Add Raster Layer option in the Quantum GIS (QGIS) desktop application interface. (From QGIS.)

2. Next adjust the band combination displayed in QGIS by right-clicking on the image layer (*QGIS_T31UDQ_20170526T105031_B02-4_8*) in the Layers Panel, found on the lower-left of the QGIS interface, and then selecting **Properties**. In the Properties dialog window, notice the current bands being displayed are Band 1 (red), Band 2 (green), and Band 3 (blue).

Remember with the Sentinel-2 MSI sensor, a TCC (originally achieved by a 4, 3, 2 band combination) would now be achieved by a creating a 3 (red), 2 (green), 1 (blue) band combination from the multi-band image since the original band 1 was not included in the layer stack.

Change the band combination displayed in QGIS to 3 (red), 2 (green), 1 (blue) to achieve natural color, or TCC image display. Click on "**Apply**" and inspect the image layer (Figure 9.36).

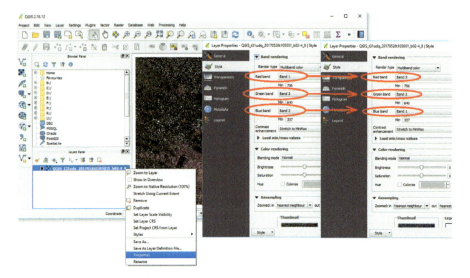

FIGURE 9.36
Adjustment of image band combination displayed in QGIS to achieve natural color, or TCC image display (3-red, 2-green, 1-blue). (From QGIS.)

3. Next create a FCC display by assigning a band combination of 4 (red), 3 (green), 2 (blue) and select "**Apply.**"

Remember in the Sentinel-2 multi-band image created in QGIS, only bands 2, 3, 4, and 8 were included in the original layer stack (or *"image merge"* as it is referred to in QGIS). The layer stack operation renamed these bands as 1, 2, 3, and 4. This means bands 8 (red), 4 (green), and 3 (blue) are being displayed for the FCC. Next click on "**OK**" to dismiss the Layer Properties window (Figure 9.37).

Unsupervised Classification 201

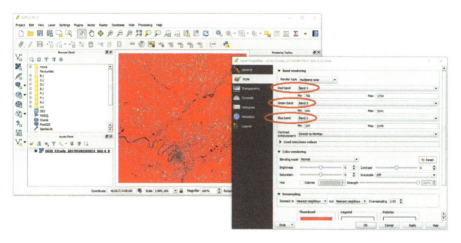

FIGURE 9.37
Adjustment of image band combination displayed in QGIS to achieve a FCC image display (4-red, 3-green, 2-blue). (From QGIS.)

4. Next create an image subset (raster clip) similar to the area in the Sentinel-2 subset image created in the previous section (t31udq_20170526t105031_b02-4_8subset.img).

   In the Processing Toolbox (displayed on the right-side of the QGIS interface), type "**clip raster**" into the search bar. If the Processing Toolbox is not displayed, click on **Processing** in the top file menu, and then **Toolbox**.

   Next select, **GDAL/OGR | [GDAL] Extraction | Clip raster by extent**. GDAL refers to the Geospatial Data Abstraction Library for raster and vector geospatial data formats. Whereas, OGR refers to the OpenGIS Simple Features Reference Implementation, which allows some writing and reading of vector file formats and is a part of the GDAL library.

   In the Clip raster by extent window that opens, click on the ellipsis button to the right of the **Clipping extent** field and select "**Select extent on canvas**."

   Next click and drag the mouse across the region you wish to subset. Ensure the "**Open output file after running algorithm**" option is checked and click on "**Run**" (Figure 9.38).

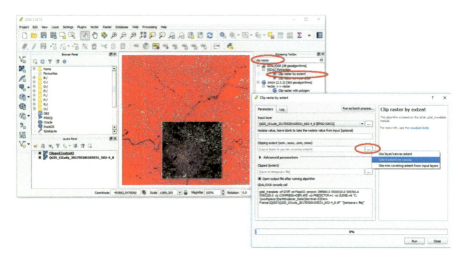

FIGURE 9.38
Raster clip (image subset) options in QGIS. (From QGIS.)

5. Save the new layer by right-clicking on the "**Clipped (extent)**" layer in the Layers Panel, select "**Save As...**" and name the new layer "**Clip_QGIS_T31UDQ_20170526T105031_subset.tif**." Ensure the "**Add saved file to map**" option is checked. Then select "**OK**."

6. Remove the "**Clipped (extent)**" and "**QGIS_T31UDQ_20170526T 105031_B02-4_8**" layers in the Layers Panel by **right-clicking** on each layer and selecting "**Remove**."

7. Right-click on the "**Clip_QGIS_T31UDQ_20170526T105031_ subset.tif**" in the Layers Panel and select "Zoom to layer" to expand the image within the display. Finally, apply a FCC display by assigning a band combination of 4 (red), 3 (green), 2 (blue) and select "**Apply**."

8. To begin an unsupervised classification, go to **Processing Toolbox** (on the right-side of the QGIS interface) | **Orfeo Toolbox (image analysis) | Learning | Unsupervised KMeans image classification**.

9. In the Unsupervised KMeans image classification window, complete the following parameters; Number of classes: **40**, Output Image: **GIS_paris_may2017_unsupervised.tif**. Ensure that the "**Open output file after running algorithm**" is checked, and select "**Run**" to begin the classification (Figure 9.39).

Unsupervised Classification 203

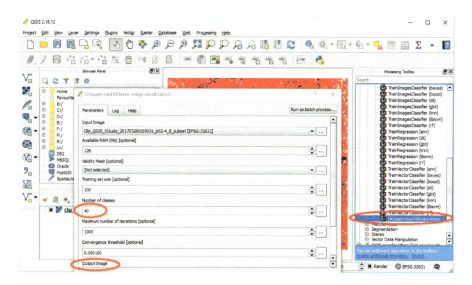

FIGURE 9.39
Unsupervised KMeans image classification options in QGIS. (From QGIS.)

10. Next, right-click on the new "**Output Image**" in the Layers Panel and select **Properties**. In the Layer Properties window, complete the following parameters; Render type: **Singleband pseudocolor**, Interpolation: **Discrete**, Color: **Greys**, Min: **0**, Max: **40**, Mode: **Equal interval**, and Classes: **40 (or higher if you desire)**. Then select "**Apply**." Next, click on the **Plus** icon next to the "**Output Image**" in the Layers Panel to see the class value colors (Figure 9.40).

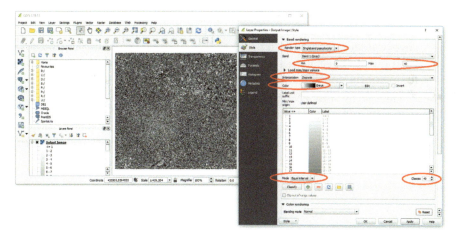

FIGURE 9.40
Output Image Style options in the Layer Properties dialog window. (From QGIS.)

11. In the Layer Properties window, double-click on the first value color. In the Change color window that opens, select an Alarm Color (such as yellow or red) for the first-class value and select "**OK**," and "**Apply**" in the Layer Properties window. Notice that the corresponding color on the "**Output Image**" in the QGIS display will update based on the color selection (Figure 9.41).

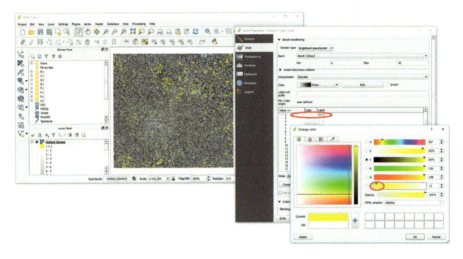

FIGURE 9.41
Evaluation of each output class value generated by the unsupervised classification algorithm. (From QGIS.)

Next begin "**working**" through each output class value generated by the unsupervised classification algorithm by turning the color of each successive output class (value 2, value 3, etc.) initially to the alarm color, and then to the designated color for each of the five land cover categories identified previously. Continue "**working**" through each output class value until all 40 class values have been assigned to one of the five land cover categories.

Decide on the appropriate land cover category for each class value, then double-click on the value in the Label column and enter the land cover name (Figure 9.42).

Unsupervised Classification 205

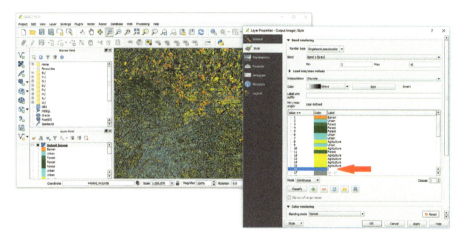

FIGURE 9.42
Selection of colors to represent the five land cover categories. (From QGIS.)

For the unsupervised classification, you should try to identify the following five classes, along with the specified color and definition scheme:

Land Cover Category and Color to apply

　　Class 1: Water—Blue

　　Class 2: Urban/Developed Land—Cyan

　　Class 3: Agriculture/Herbaceous Vegetation—Yellow

　　Class 4: Forested Vegetation—Dark Green

　　Class 5: Barren Land—Orange

See Table 9.1 for a description of each land cover category.

12. You may also find it helpful to use the Zoom In and Zoom Out tools. Also, you may find it useful to add another instance of the "**Clip_QGIS_T31UDQ_20170526T105031_subset.tif**" as a TCC display to the QGIS Layers Panel, and click and drag the layer below the "Output Image." By checking and unchecking the "**x**" icon next to each layer you can toggle them off and on.

13. The final unsupervised classification should resemble something similar to the following figure. Remember to click on Project | Save As ... to save the data as a QGIS project (*.qgs) (Figure 9.43).

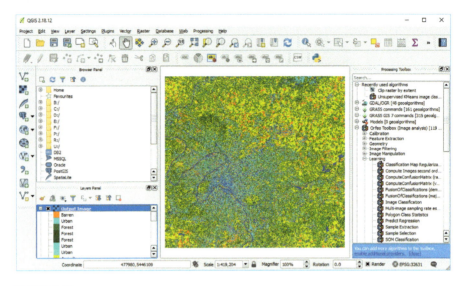

FIGURE 9.43
QGIS unsupervised classification result. (From QGIS.)

NOTE: Only **40 output classes for the unsupervised classification** were specified. No classification is ever 100% accurate. However, remember the classification results may be "**GREATLY IMPROVED**" by specifying a **higher** number of output classes for each desired land cover category that you would like to achieve in the thematic map. Additionally, it is a good idea to use the spectral profile tool to examine representative classes throughout the entire raw image to get an idea of how similar or variable the spectral values may be for each class.

Review Questions

1. What does the term classification mean in relation to the conversion of raw imagery into a thematic image?
2. What are the two traditional image classification methods called?
3. Explain, in general terms, the process of unsupervised classification.
4. Explain, in general terms, the Iterative Self-Organizing Data Analysis Technique (ISODATA) clustering algorithm typically used in the generation of an unsupervised classification.
5. The k-means approach represents a second clustering algorithm that also may be used in the generation of an unsupervised classification. Discuss the operations of this approach.

10

Supervised Classification

Overview

In the previous chapter, image classification as the operation of grouping or sorting image pixels, typically based on the pixel spectral values, into categories (or themes) was discussed. Also discussed was the process of unsupervised classification, which requires minimal image analyst input to develop a thematic map output. The second of the two traditional image classification methods is known as a supervised classification. The exercise associated with this chapter's application focuses on developing a supervised classification. This exercise will explore creating a supervised classification.

Supervised Classification: A supervised classification process generally consists of three stages performed by the analyst. These stages include training, classification, and the final output. The training stage consists of the initial stage of a supervised classification. In this stage, the analyst inspects the image to be classified and uses knowledge of the area (collected from field visits, reference maps or photos, or other higher quality and higher resolution data) to collect training sites within the imagery that represent the corresponding areas on the ground. The training sites may be collected in the form of delineated polygons or representative pixels that the software will use to develop a multiband classification based on spectral relationships scattered from the sites. Each training site should represent a homogeneous and contiguous grouping of pixels within an individual category of interest. The number of training sites collected from the imagery should also capture the amount of variability contained within the category of interest as identified across the entire image data. Also, each training site's category of interest should be randomly or systematically distributed throughout the entire image data (Khorram et al. 2016b).

The classification process represents the second stage of supervised classification. This process uses statistical algorithms to analyze the spectral bands of the imagery and determines how closely each relates to the identified training samples representing the categories of interest throughout the entire image data. The statistical algorithms represent the most widely used algorithms in the classification process. These include the minimum distance to means, parallelepiped, and maximum likelihood.

The minimum distance to means and parallelepiped algorithms tends to be less processor intense than the maximum likelihood algorithm. In the minimum distance to means classification, the algorithm first determines the mean spectral values in each band for each training sample category. Then spectral ranges, or domains, are computed for each band. Each land cover category, within a well identified spectral space, will be assigned to a category based on the mean vector within the category of interest. Unassigned pixels will be assigned to a category based on the minimum distance between the unassigned pixel's value and the mean vector within range of the closest category of interest. The minimum distance to means classification tends to work well for image data that consists of fewer intended category classes of interest, or imagery with classes that are more homogenous. Image data with higher degrees of variance in spectral response data between classes may produce decreased classification accuracies (Lillesand et al. 2008).

The parallelepiped classification algorithm overcomes some the limitations of the minimum distance to means algorithm, with respect to dealing with higher degrees of spectral variance across an image dataset. The parallelepiped classification algorithm establishes a spectral range within each training site that represents the intended category of interest. The spectral range represents groupings of the lowest and highest pixel values and is arranged in a rectangular, or stepped rectangular scattergrams. These groupings are also called parallelepipeds, which is where the algorithm gets its name. The parallelepipeds are established for each category within the multiple bands of the image dataset. However, covariance, or the tendency of spectral values to vary similarly in two or more bands, become a problem as this may decrease the interdependency of bands to each other (Lillesand et al. 2008).

The maximum likelihood classification algorithm assumes that the statistical distribution of training data, for each class and in each band, will follow a normal, Gaussian distribution (Blaisdell 1993). Given a normal statistical distribution, it is assumed that the mean vector and covariance matrix can be used to identify the statistical probability, or likelihood, of spectral values in each category belonging to one class or the other. Based on the probability density function an unknown or unassigned, class pixel can then be assigned the category of the highest probability value. Equiprobability contours are established based on their sensitivity to the covariance between image bands within the datasets (Lillesand et al. 2008). The maximum

Supervised Classification 209

likelihood classification algorithm represents the most popular method of the three classification algorithms. However, this method is generally more resource-intensive than the minimum distance to means and parallelepiped algorithms (Hester et al. 2010, Hord 1982).

The last stage of the supervised classification process is known as the final output stage. The final output may be produced as a thematic, or classification map, representing each category of interest developed originally from the training samples. The thematic data may further be analyzed to determine land use and land cover areas for each defined category. Additionally, thematic data be imported into other geospatial and statistical tools for further mapping and modeling. Classified data products may also be used for visualizing and presenting land use and land cover analysis results, or even modeled with other temporal land cover data to develop change analysis models. The thematic classification data, along with the accompanying statistical parameters, may also be used to develop an assessment of accuracy to determine an estimate of the thematic classification accuracy.

Supervised Classification Application

Learning Objectives

1. To gain an understanding of the general procedures for image classification using the unsupervised image analysis process to classify land use and land cover.
2. To gain an understanding of the general procedures for image classification using the supervised image analysis process to classify land use and land cover.

Data Required: This exercise will demonstrate a supervised classification approach on a Landsat 8 Operational Land Imager/Thermal Infrared Sensor (OLI/TIRS) image of the town Orvieto, in the Province of Terni, Italy. This image will be downloaded, along with a Sentinel-2 reference image, that will be used for a visual comparison of classification agreement. The images can be downloaded from the United States Geological Survey (USGS) EarthExplorer (http://earthexplorer.usgs.gov/) website as previously described in Chapter 1.

210　　　*Image Processing and Data Analysis with ERDAS IMAGINE®*

Obtaining the Required Data (Review)

From the USGS EarthExplorer website, the Landsat 8 image (**LC08_L1TP_191030_20170706_20170716_01_T1**, Acquisition Date: 2017/07/06) and a Sentinel-2 image (**L1C_T32TQN_A010601_20170703T101041**, Acquisition Date: 2017/07/03) for the town of Orvieto, Italy area will be downloaded.

1. In the Address/Place field, type in the word *"Orvieto"* and click on the "**Show**" button below the box. Next, click on the **05018 Orvieto, Province of Terni, Italy** link under the Address/Place heading to show the location on the map and add coordinates to the Area of Interest Control.

2. In the "**Search from**" boxes in the Search Criteria section enter the following dates: **05/01/2017** to **08/31/2017**. Next select "**Data Sets**" below Date Range options.

3. Select "**Landsat Archive | L8 OLI/TIRS C1 Level-1**" and "**Sentinel | Sentinel-2**" on the Data Sets tab.

4. Next select "**Results**" at the bottom of Data Sets tab. Now inspect the available imagery.

5. Select the download icons for the following images and unzip them as previously described in Chapter 1.

   Landsat 8 OLI/TIRS L8 C1 Level-1 image

   LC08_L1TP_191030_20170706_20170716_01_T1

   Acquisition Date: 2017/07/06

   Sentinel-2 MSI image

   L1C_T32TQN_A010601_20170703T101041

   Acquisition Date: 2017/07/03

6. After unzipping each image, you will need to perform a layer stack within ERDAS IMAGINE® as previously described in Chapter 1.

7. Next create image subsets of Orvieto, Italy for both images as previously described in Chapter 1 (Figure 10.1).

Supervised Classification 211

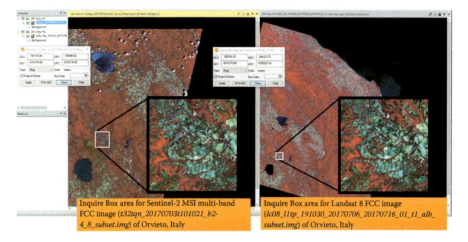

FIGURE 10.1
Image subset locations for the multi-band Sentinel-2 MSI image (L1C_T32TQN_A010601_20170703T101041: Acquisition Date: 2017/07/03; on left) and the multi-band Landsat 8 OLI/TIRS image (LC08_L1TP_191030_20170706_20170716_01_T1: Acquisition Date: 2017/07/06; on right). (From ERDAS IMAGINE®/Hexagon Geospatial.)

8. These images represent an input classification image for the supervised classification (Landsat 8, 30 m resolution) and a reference image (Sentinel-2, 10 m resolution) image that will be used for a visual comparison of classification agreement (Figure 10.2).

FIGURE 10.2
Subset images sentinel-2 MSI (left) and Landsat 8 OLI/TIRS (right) images of Orvieto, Italy. (From ERDAS IMAGINE®/Hexagon Geospatial.)

Initiating the Supervised Classification

This exercise will attempt to identify similar land cover classes performed on the unsupervised classification for Paris, France in the previous chapter (Chapter 9).

NOTE: When opening the image in ERDAS IMAGINE, check the "**Raster Options**" tab to ensure that a standard false color composite band combination (5,4,3) is applied to the display of the Landsat 8 subset image of **Orvieto, Italy**. Use the **Metadata** icon [ⓘ] from the Home tab to obtain information about the image file. Take notice of the appearance of specific ground features displayed on the data using the standard false color composite band combination (e.g., the color of vegetation within the image), and whether or not the composite appears to enhance any particular feature.

For the supervised classification you should try to identify the following six classes ("unclassified" will be included by default as a seventh class), along with the specified color and definition scheme:

Land Cover Category and Color to apply

Class 0: Unclassified—Black

Class 1: Water—Blue

Class 2: Urban/Developed Land—Cyan

Class 3: Agriculture/Herbaceous Vegetation—Yellow

Class 4: Evergreen Vegetation—Dark Green

Class 5: Deciduous Vegetation—Green

Class 6: Barren Land—Orange (Table 10.1)

Supervised Classification

TABLE 10.1

Land Cover Classification Class and Class Descriptions for the Supervised Classification of the Town Orvieto, in the Province of Terni, Italy

Land Cover Classification	Classification Description
Urban/Development	Contained industrial, commercial, and private building, and the associated parking lots, a mixture of residential buildings, streets, lawns, trees, isolated residential structures or buildings surrounded by larger vegetative land covers, major road and rail networks outside of the predominant residential areas, large homogeneous impervious surfaces, including parking structures, large office buildings, and residential housing developments containing clusters of cul-de-sacs.
Agriculture/Herbaceous Vegetation	Contained agricultural field, grass fields, and urban grasses. Characteristics of this class include large agricultural field, mowed/maintained lawns, fields, and vegetated road medians.
Evergreen Vegetation	Contained large homogeneous vegetative land covers of trees or shrubs that keep their leaves throughout the winter (mostly coniferous species).
Deciduous Vegetation	Contained large homogeneous vegetative land covers of trees or shrubs that lost their leaves during the winter (mostly hardwood species).
Barren Land	Contained areas or fields with little or no vegetation. Characteristics of this class include fallow agricultural fields, bare sediment or soil areas, and areas cleared of vegetation for construction.
Water	Contained lakes and small ponds and all other natural and artificial surface waters outside of the USGS National Hydrography Dataset shapefile.
Unclassified	Contained areas of unclassified or misclassified data.

Source: Hester D.B. et al., *Photogramm. Eng. Remote Sensing*, 74, 463–471, 2008.

Supervised Classification Application Approach—Creating a Signature File

To perform a supervised classification on the Landsat 8 subset image of **Orvieto, Italy**, you will first need to create a signature file. A signature file has information on the characteristic spectral response of each category of interest. To begin the supervised classification, select the following options:

1. Add the Landsat 8 subset image of Orvieto to the 2D View window as a False Color Composite (FCC) image display (5, 4, 3).

2. Select **Raster | Supervised** within the Classification category grouping **| Signature Editor**. This opens the Signature Editor tool window.

   With the subset image displayed in a viewer, the ERDAS IMAGINE's Area of Interests (AOI) tools will be used to create polygons and delineate homogeneous areas representing the categories of interest. This is accomplished by the following steps:

3. Select **File | New | 2D View #1 New Options | AOI Layer** (Figure 10.3).

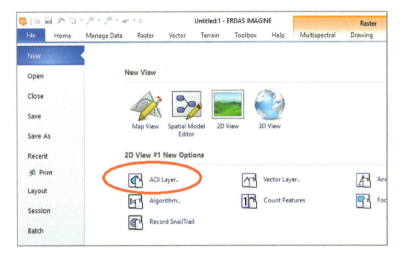

FIGURE 10.3
Add AOI layer option. (From ERDAS IMAGINE®/Hexagon Geospatial.)

4. In the main menu with the newly created AOI layer active in the Table of Contents (this should be highlighted/active by default), click on the "**Drawing**" tab in the AOI menu tab grouping.

   NOTE: Again, ensure the new AOI layer is the active layer in Contents list (Figure 10.4).

Supervised Classification 215

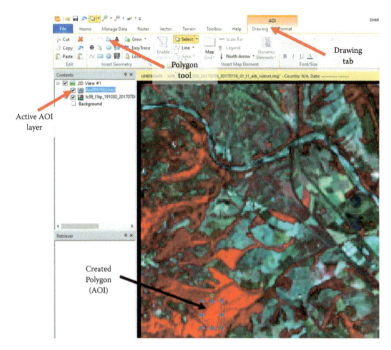

FIGURE 10.4
Creating training site polygons in the active AOI layer. (From ERDAS IMAGINE®/Hexagon Geospatial.)

5. Use the **Polygon tool** icon (within the **Insert Geometry** category) grouping to create polygons for homogeneous areas.
6. Click once to create each vertex of the polygon and double-click to complete.

    As you create each polygon, in the Signature Editor tool window add the newly captured signature to the signature editor by clicking the "**Create New Signature(s) from AOI**" tool. This is accomplished by the following steps:

7. Select **Edit | Add** in the signature editor, or the click on the "**Create New Signature from AOI**" button in the signature editor (Figure 10.5).

FIGURE 10.5
Create New Signature form AOI. (From ERDAS IMAGINE®/Hexagon Geospatial.)

NOTE: Make sure you also edit the name of the class to the category you are delineating to match the names the land cover categories of interest. Each added signature name must be unique (i.e., Forest1, Forest2, Forest3, etc.). Additionally, change the color of each class signature to the representative color specified earlier for each class category (water—blue, urban—cyan, agriculture—yellow, etc.). Be sure to continue adding AOIs for all six land cover classes (Figure 10.6).

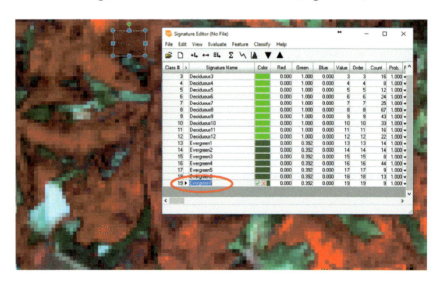

FIGURE 10.6
Edit the name of the class of polygon training site category being delineated to match the names the land cover categories of interest. Each added signature name must be unique (i.e., Forest1, Forest2, Forest3, etc.). (From ERDAS IMAGINE®/Hexagon Geospatial.)

Supervised Classification 217

8. Once you have completed delineating several representative poly-gons for each category represented in the image (more is **ALWAYS** better than a few), **save** the signature file by selecting **File | Save** in the Signature Editor tool window. Also, be sure to save your AOI layer (**Right-Click on the AOI file** in the Contents Legend | **Save As**).

9. Next, select **Classify | Supervised Classification** in the Signature Editor window and run the classifier with the signature file you created. Specify an Output File name in the supervised classification window (see note in the following about naming conventions) and then click OK to run the classification. A Process List window should appear and begin the procedure. Once processing is complete, close the process list window.

NOTE: Should ERDAS IMAGINE crash, or experience an error and close, when running the Supervised Classification from the Signature Editor window, you may also run the Supervised Classification from the tabbed ribbons menu (**Raster** tab | **Supervised** in the Classification menu grouping | **Supervised Classification**).

NOTE: When naming the output file, it is a good idea to use a naming convention that allows you to easily recognize the origin of the file and any subsequent operations made to the original file. Also, it is a good idea to document these changes, similar to Table 10.2.

TABLE 10.2

Image File Naming Conventions. Type in Bold Represents the Subsequent Modification of the Original Image File

Operation	File Name	Type of Operation
Original file	lc08_l1tp_191030_20170706_201 70716_01_t1.img	Original file
subsequent operation 1	lc08_l1tp_191030_20170706_201 70716_01_t1_**alb**.img	Multi-band all bands (ALB)
subsequent operation 2	**Orvieto**_alb.img	Layer Save As (renamed Orvieto for convenience)
subsequent operation 3	Orvieto_alb_**subset**.img	Image subset
subsequent operation 4	Orvieto_alb_subset_**supclass**. img	Supervised classification
etc.	...	...

10. Once you have completed the supervised classification, close the Signature Editor tool window and remove the AOI layer from the Contents Legend (leave the Landsat 8 subset input image open in the 2D View window). Next, display the newly created supervised

classification output image in a viewer by adding the raster to the Contents Legend. The supervised classification output image is now known as a "**Thematic Image**" (Figure 10.7).

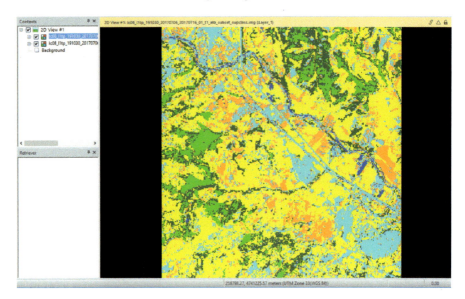

FIGURE 10.7
Completed supervised classification, known as a classified, or thematic image. (From ERDAS IMAGINE®/Hexagon Geospatial.)

Tips for Creating the Supervised Classification

When capturing signatures, it is a good idea to keep your AOIs as small as possible to minimize (pixel spectral) variability within the polygon. Also try to capture as many polygons as you think is necessary to fully represent the class category throughout the entire image (for instance, Do **not** capture all the polygons from the **same area**, but be sure to spread the AOIs out).

You may also use the zoom tools to, zoom in and out, to help you determine the correct class category. Also, you may want to change the band combinations from FCC, to True Color Composite (TCC), and back to FCC to help you identify the correct class categories. Typically, it is not a good practice to use reference imagery, as this imagery will be used for a visual comparison of classification agreement and perform an accuracy assessment later.

To determine deciduous vegetation from coniferous, or evergreen vegetation, when displaying an image as a FCC display, remember that healthy green vegetation will reflect strongly in the near infrared (shortwave IR) of the electromagnetic spectrum (EMS). Due to the leaf structure of most evergreen trees, in a FCC display, will typically appear darker red in comparison to the deciduous trees. The deciduous trees, generally having broader leaves,

Supervised Classification

reflect more near infrared light creating a brighter red throughout the image when compared to evergreen trees (Figure 10.8).

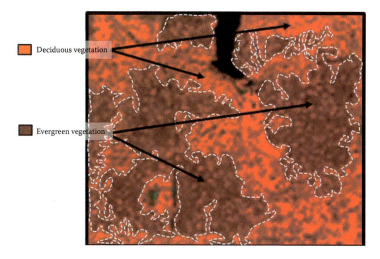

FIGURE 10.8
Comparison of deciduous (lighter red) and evergreen vegetation (darker red) in a FCC display. (From ERDAS IMAGINE®/Hexagon Geospatial.)

An additional tip in creating the supervised classification is to use the Spectral Profile tool. As discussed in Chapter 6, the Spectral Profile tool may be very helpful in determining how similar the signature of the AOI capture areas are to each other. Ideally, you would want all the individual class AOIs to be as close to each other as possible to ensure you have captured well-defined classes.

Finally, keep in mind, classification is not an exact science. *Picking pixels* out of an image definitely takes a lot of patience and practice. As a matter of fact, the more time that you spend with the imagery to be classified, the better the overall classification becomes. With practice, as well as understanding, the rules of how light interacts with the features on the surface of the earth, your judgment will become increasingly better.

Compare Using Swipe Tool

1. To visually compare your classification results (thematic image), use your swipe tool. You should still have the Landsat 8 subset image and the newly created thematic image open in the same 2D View window.
2. Rearrange the image order in the Contents Legend by dragging the thematic image to the top of the list, so that the Landsat 8 subset image is below the thematic image (Figure 10.9).

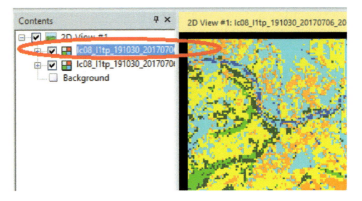

FIGURE 10.9
Rearrange the image layer order in Contents Legend by dragging the thematic layer to the top of the list (shown in 2D View #1). (From ERDAS IMAGINE®/Hexagon Geospatial.)

3. Open the swipe tool (**Home** tab | **Swipe** within the View category grouping) to compare the land cover classification for areas of agreement and disagreement (Figure 10.10).

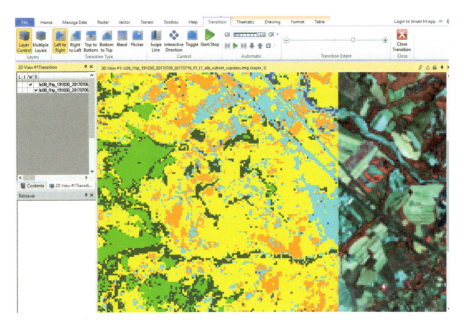

FIGURE 10.10
Swipe tool comparison of the thematic image and Landsat 8 subset image. (From ERDAS IMAGINE®/Hexagon Geospatial.)

Supervised Classification 221

a. Next, compare the thematic image to the higher resolution Sentinel-2 subset image of the same area (this is the reference image). To do so, first close the swipe transition tool. Then, from the Home tab | Contents Legend, remove the Landsat 8 subset image (**Right-Click | Remove Layer**).
b. Next, add the Sentinel-2 subset image as a FCC (4, 3, 2) to the same 2D View as the thematic image and rearrange the image order in the Contents Legend by dragging the thematic image to the top of the list, so that the Sentinel-2 subset image is below the thematic image.
c. Open the swipe tool (**Home** tab | **Swipe** within the View category grouping) to compare the land cover classification for areas of agreement and disagreement (Figure 10.11).

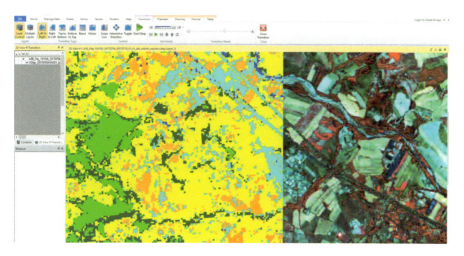

FIGURE 10.11
Swipe tool comparison of the thematic image and Sentinel-2 subset image. (From ERDAS IMAGINE®/Hexagon Geospatial.)

Compare Using Image Difference Operation (Change Detection)

Image classifications may also be examined using a change detection operation. Change detection differs from the visual inspection, which we examined earlier using the swipe tool. This operation requires an additional classified image that is used to compare changes occurring either before or after the original image date.

A common change detection operation, which is used to compute the differences that occur between the two dates of the classified images, is known as an *Image Difference* operation. This operation subtracts the *After Image* (the later dated, classified image) from the *Before Image* (the earlier dated, classified image). The output produces two types of image files. The first output image file is a grayscale, image difference file that represents the changes in brightness values over time between the *before* and *after* classified input images. The second output image file highlights the areas where classes have increased in change (class area) from one image to another, decreased in change from one image to another, or remained unchanged from one image to another. The highlighted change areas are typically specified by a user-determined threshold value or percentage (such as +/− 10% change).

1. To create an image difference change detection, first use the supervised classification procedures described earlier to create a supervised classification from the Sentinel-2 image (**L1C_T32TQN_A010601_20170703T101041**, Acquisition Date: 2017/07/03) for the town of Orvieto, Italy. This image was previously downloaded from the USGS EarthExplorer website. After repeating the procedures described earlier to complete a supervised classification, this resultant image classification will represent the *After Image*.

2. Next, download a second Sentinel-2 image from the USGS EarthExplorer website for the town of Orvieto, Italy that corresponds in area to an earlier date of the image previously downloaded (**2A_OPER_MSI_L1C_TL_EPA__20150704T102427_20160809T015434_A000162_T32TQN_N02_04_01**, Acquisition Date: 2015/07/04). After completing a layer stack operation on this image, repeat the supervised classification procedures described earlier using the identical six land cover categories that were used in the previous date Sentinel-2 image. This image will present the *Before Image* as it was captured approximately two years before the *After Image*.

   NOTE: You may wish to subset the two images into identical corres-ponding image areas to reduce your classification effort and focus your comparison on only the area within each image that you are interested in exploring the changes occurring from one date to the next.

3. Display the results of your two supervised classifications in separate 2D View windows. You may also wish to perform a thematic recode

Supervised Classification   223

on each completed classification to represent the six land cover categories (1-Water, 2-Urban, 3-Agriculture/Herbaceous Vegetation, 4-Evergreen Vegetation, 5-Deciduous Vegetation, 6-Barren Land) (Figure 10.12).

FIGURE 10.12
Supervised classification results of the two Orvieto, Italy, Sentinel-2 MSI images. Image orvieto_b2348_july32017.img (L1C_T32TQN_A010601_20170703T101041, Acquisition Date: 2017/07/03) represents the *Before Image* and the image orvieto_b2348_july42015.img (2A_OPER_MSI_L1C_TL_EPA__20150704T102427_20160809T015434_A000162_T32TQN_N02_04_01, Acquisition Date: 2015/07/04) represents the *After Image*. (From ERDAS IMAGINE®/Hexagon Geospatial.)

4. Next, select **Raster | Change Detection Tools | Image Difference**. In the Change Detection Using Image Difference dialog window that opens, add the 2015 Orvieto, Italy image as the *Before Image* and the 2017 Orvieto, Italy image as *After Image*. Also enter the file names "Image Difference File" and "Highlight Change File." Select the "As Percent" option to compute the change value thresholds (increases and decreases) as percentage values. You may also select colors of your preference to represent these values (Figure 10.13).

224 *Image Processing and Data Analysis with ERDAS IMAGINE®*

FIGURE 10.13
Change Detection Using Image Difference dialog window with 2015 Orvieto, Italy image as the *Before Image* and the 2017 Orvieto, Italy image as *After Image*. (From ERDAS IMAGINE®/Hexagon Geospatial.)

5. Display the "Image Difference File" and the "Highlight Change File" results in separate 2D View windows (Figure 10.14).

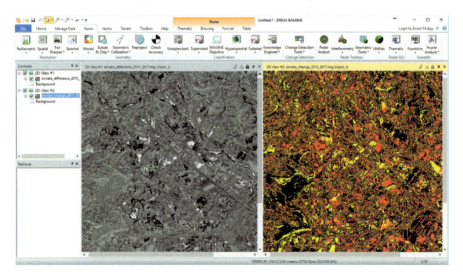

FIGURE 10.14
Image difference results. Image Difference File (2D View #1 on left) represents a represents the changes in brightness values over time between the before and after classified input images. Highlight Change File (2D View #2 on right) represents areas that have increased or decreased based on the user-specified threshold (Yellow—increases more than 10%, Red—decreases more than 10%). (From ERDAS IMAGINE®/Hexagon Geospatial.)

Supervised Classification

6. Select **Raster | Thematic | Summary Report of Matrix**. This operation compares the class values between two input files by designating one file as the zone file and the other file as the class file. In the Summary dialog window that opens, add the 2015 Orvieto, Italy image as the "Input Zone File" and the 2017 Orvieto, Italy image as "Input Class File." The Input Zone File (or zone layer) uses each class as an analysis category to compute statistics of pixel count, percentage, and area in common for each category based on the class occurrences within the Input Class File. Select the "Interactive (CellArray)" option and then "OK" to display the Summary By Zone dialog (Figure 10.15).

FIGURE 10.15
ERDAS Summary Report of Matrix based on statistics computed between input zones (or classes used as analysis categories) from 2015 Orvieto, Italy image for each category based on the class occurrences within the input class file (2017 Orvieto, Italy image). (From ERDAS IMAGINE®/Hexagon Geospatial.)

NOTE: The "Output Report Option" may also be selected in the Summary dialog window to save a matrix summary report as a text file by clicking the "Output Report Only" instead of the "Interactive (CellArray)." The summary report text file generated by ERDAS IMAGINE will require some effort to reformat it from its original format for the ease of interpretation of the report's results. A common way of displaying the results resembles a land-cover transition matrix. The land-cover transition matrix may be used to compare class value areas between the 2015 Orvieto, Italy image (*Before Image*) and 2017 Orvieto, Italy image (*After Image*). Figure 10.16 shows the summary output report that has been reformatted into a land-cover

Pixel counts in common

2017 \ 2015	Unclassified	Water	Urban	Agriculture/Herbaceous vegetation	Evergreen vegetation	Deciduous vegetation	Barren land
Unclassified	84	2	3	82	11	35	0
Water	1	1971	170	229	53	118	48
Urban	120	3238	45316	87562	10692	7001	18378
Agriculture/Herbaceous Vegetation	174	2331	52225	153891	28482	18395	12274
Evergreen vegetation	89	2323	1019	19933	48586	12461	129
Deciduous vegetation	43	159	240	9088	21580	58217	22
Barren land	4	112	6534	7970	566	589	7225
TOTAL	431	10134	105504	278673	109959	96781	38076

Percentages in common

2017 \ 2015	Unclassified	Water	Urban	Agriculture/Herbaceous vegetation	Evergreen vegetation	Deciduous vegetation	Barren land
Unclassified	0	0	0	0	0	0	0
Water	0.23	19.45	0.16	0.08	0.05	0.12	0.13
Urban	27.84	31.95	42.95	31.42	9.72	7.23	48.27
Agriculture/Herbaceous Vegetation	40.37	23	49.5	55.22	25.9	19.01	32.24
Evergreen vegetation	20.65	22.92	0.97	7.15	44.19	12.88	0.34
Deciduous vegetation	9.98	1.57	0.23	3.62	19.63	60.15	0.6
Barren land	0.93	1.11	6.19	2.86	0.51	0.61	18.9
TOTAL	100	100	100	100	100	100	100

Hectare in common

2017 \ 2015	Unclassified	Water	Urban	Agriculture/Herbaceous vegetation	Evergreen vegetation	Deciduous vegetation	Barren land
Unclassified	0.84	0.02	0.3	0.82	0.11	0.35	0
Water	0.01	19.71	1.7	2.29	0.53	1.18	0.48
Urban	1.2	32.38	453.16	875.62	106.92	70.01	183.78
Agriculture/Herbaceous Vegetation	1.74	23.31	522.25	1538.91	284.82	183.95	122.74
Evergreen vegetation	0.89	23.23	10.19	199.33	45.86	124.61	1.29
Deciduous vegetation	0.43	1.59	2.4	90.88	215.8	582.17	0.22
Barren land	0.04	1.12	65.34	79.7	5.66	5.89	72.25
TOTAL	4.31	101.34	1055.04	2786.73	1099.59	967.81	380.76

FIGURE 10.16

ERDAS Summary Report of Matrix—Output Report Option. The output report text file may be reformatted as a land-cover transition matrix to compare class value areas common between the 2015 Orvieto, Italy image (Before Image) and 2017 Orvieto, Italy image (*After Image*). The numbers in the diagonal represent values that have not changed between the images. The matrix values outside of the diagonal are interpreted as change between the *Before* and *After* images.

Supervised Classification 227

transition matrix. The rows represent class values (pixel counts, percentages, and hectare) found in the 2015 Orvieto image and found in the 2017 Orvieto (columns). The numbers in the diagonal represent values that have not changed between the images. The matrix cell values outside of the diagonal are interpreted as change between the *Before* and *After* images. For example, in evaluating percentages in common matrix, approximately 10% (or 9.72%) of the Urban class values in the 2017 Orvieto image (*After image*) were found in common with the Evergreen Vegetation class values in the 2015 Orvieto image (*Before image*). This is interpreted as a 10% (or 9.72%) change from the Evergreen Vegetation class values in the 2015 image to Urban class values in the 2017 image.

The land-cover transition matrix is also useful for determining where land cover conversions were most prominent (such as the conversion of agriculture to urban land), and whether or not all land cover categories changed as might be expected (e.g., agriculture land changed primarily to urban; barren land classes changed primarily to agriculture, urban, or natural vegetation classes; and natural vegetation classes changed primarily to urban and/or other natural vegetation classes). For more information on the development and interpretation of land-cover transition matrices from remotely sensed data, please see Nelson et al. (2002).

7. Finally, select **Raster | Thematic | Matrix Union**. This operation creates an output file containing class values that overlap within the input files. In the Matrix Union dialog window that opens, add the 2015 Orvieto, Italy image as the "Thematic Image/Vector #1" image and the 2017 Orvieto, Italy image as "Thematic Image/Vector #2." Name the "Output File" and select the "Intersection" option to perform the algorithm on only the image area that both files have in common. The attribute table from the result of the Matrix Union output image displays the *Before Image* class value in "orvieto_b2348_2015 value" column and the *After Image* class value in "orvieto_b2348_2017 value" column (Figure 10.17).

228   *Image Processing and Data Analysis with ERDAS IMAGINE®*

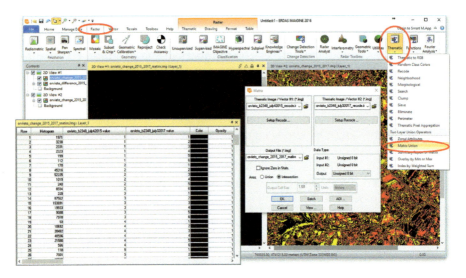

FIGURE 10.17
ERDAS Matrix Union result displayed in 2D View #1 (orvieto_change_2015_2017_matirx.img) and attribute table. Attribute table displays the *Before Image* class value in "orvieto_b2348_2015 value" column and the *After Image* class value in "orvieto_b2348_2017 value" column. (From ERDAS IMAGINE®/Hexagon Geospatial.)

Note on Image-To-Image Change Detection Comparisons

An image-to-image change detection comparison can be very powerful for examining changes occurring over time (image date 1 vs. image date 2). This information can be displayed either spatially or reported statistically based on the image class values. However, a few considerations should be noted in order to produce reliable estimates. For example, identical thematic land cover categories should be generated for both images (*Before Image* and *After Image*) during the individual classification process for each image. As well, both classified images should possess the same spatial resolution and be co-registered to the same geographical extent. Additionally, an image-to-image change detection comparison will produce more reliable estimates after individual accuracy assessments have been completed for each classified image. The accuracy assessments may be able to determine class areas that were not classified well in each image and may need to be reevaluated before attempting an image-to-image change detection comparison. Without an evaluation of the accuracy assessment for each image to gain an understanding of where incorrect or less accurate class values occur, it may be difficult to determine any real changes occurring between the *before* and *after* images. Procedures for assessing thematic classification accuracy are presented in Chapter 13.

Supervised Classification 229

Review Questions

1. Explain, in general terms, the process of a supervised classification.
2. A supervised classification generally consists of three stages performed by the analyst. Name these three stages.
3. Discuss the first stage of a supervised classification, as identified in the question earlier.
4. Discuss the second stage of a supervised classification, as identified in question #2.
5. Discuss the third stage of a supervised classification, as identified in question #2.

11

Object Based Image Analysis

Overview

The supervised or unsupervised classification methods represent a pixel-based approach. These approaches are used to generate land use/land cover classifications that are typically derived from the spectral values of the individual pixels within an image. These approaches have traditionally worked well when the identified classes throughout an image exhibit a definable spectral separation or distance. However, more recent higher-spatial resolution imagery (~1–≤1 m) has proved more challenging to use these traditional approaches due to the increase of spectral variability within the target classes (Yu et al. 2006). Object-oriented classification, image extraction, or Object Based Image Analysis (OBIA), represents a more automated method of image classification than the previously examined traditional classification approaches (unsupervised classification and supervised classification) (Shackelford and Davis 2003, Blaschke 2010). In an OBIA approach, in addition to using spectral properties, the software also uses image-based *cues* that are similar to a human's process for visual interpretation (such as color/tone, texture, size, shape, shadow, site/situation, pattern, and association). Additionally, the OBIA procedure applies classification-based objects, established from training samples, that represent the features to be classified. These objects are then defined within the software based on rules that are further used to model the individual or groups of objects based on size, shape, direction, distance, distribution throughout the image, texture, spectral pixel values, position, orientation, relationship between objects, and so on as other user-defined parameters (Khorram et al. 2016a). After each object has been identified, the OBIA procedure may then complete the classification operation by incorporating statistical algorithms such as nearest neighbor analysis, neural networks, and decision tree analyses (Herold et al. 2003, Thomas et al. 2003).

232   *Image Processing and Data Analysis with ERDAS IMAGINE®*

IMAGINE Objective is ERDAS IMAGINE's® object-oriented classification tool. In IMAGINE Objective's OBIA approach, the software processes spectral properties, as well as image-based *cues* that are similar to a human's process for visual interpretation (such as pattern, color, size, shape, etc.). Once these features are identified as an object, segmentation, or cluster, these features are then extracted from the imagery. Once all features are extracted, then they can be assigned a representative land use or land cover category for further spatial analyses. This exercise was adapted from original research developed by Mr. Kevin Bigsby at North Carolina State University. For more detailed information, visit the ERDAS IMAGINE's IMAGINE Objective website: http://www.hexagongeospatial.com/products/remote-sensing/erdas-imagine-add-ons/imagine-objective.

Object Based Image Analysis Classification Application

Learning Objectives

1. To provide an introductory to ERDAS IMAGINE's IMAGINE Objective
2. To create Project and Feature Model within IMAGINE Objective
3. To conduct a 6-class classification on example imagery using multiclass capabilities using the following classes: Wetland, Water, Barren, Forest, Cultivated, and Developed
4. To introduce a few of the common Objective tools, there are ~50 or more.

Data Required: Y1326.tif

This dataset represents a high ground spatial resolution (30 cm (~1 foot)), 4-band multispectral dataset. The Y1326.tif dataset was obtained from the U.S. Fish and Wildlife Service's Blackwater National Wildlife Refuge, Maryland.

The refuge represents an area of national importance and has been referred to as the *Everglades of the North* by the Nature Conservancy (Figure 11.1). This refuge contains one-third of Maryland's tidal wetlands and maintains an incredible amount of plant and animal diversity. To find out more about the Blackwater National Wildlife Refuge as well as to download the data used in this exercise, go to the following web address: https://www.fws.gov/refuge/Blackwater/.

Object Based Image Analysis 233

FIGURE 11.1
U.S. Fish and wildlife service's blackwater national wildlife refuge, Maryland study area land cover types.

I. Feature Project Setup Procedure

1. **Open Objective**—start ERDAS IMAGINE, search for "Objective" in help window (alternatively shown in the following; **Raster | IMAGINE Objective**) and click on Imagine Objective (Figure 11.2).

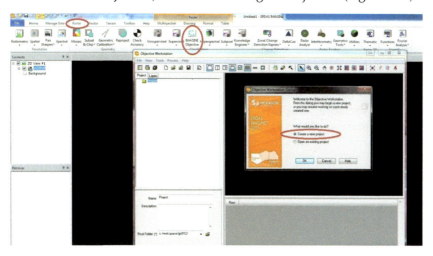

FIGURE 11.2
IMAGINE objective workstation setup. (From ERDAS IMAGINE®/Hexagon Geospatial.

2. **Create the project and Feature Model**—Next in the Objective Workstation dialog, select "Create a new project." In the graphic user interface (GUI) that pops up, fill in the appropriate information: (1) **Project Files**, (2) **Name**, (3) **Description**, (4) **Root Folder**, and (5) **New Feature Model**. Make sure all files are saved in the same directory. Make sure Setup Input Variables is checked as seen in the following image. Click on "**OK**" (Figure 11.3).

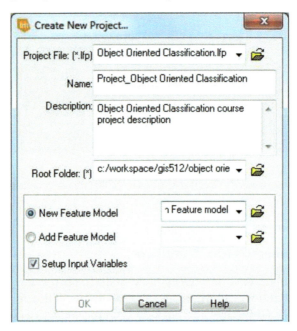

FIGURE 11.3
IMAGINE objective create new project dialog window. (From ERDAS IMAGINE®/Hexagon Geospatial.

3. **Add New Variables**—This is where you specify the imagery you want to classify. ERDAS just calls it *a variable*. Click on the "**Add New Variable**" button. After you add the variable, specify the input imagery from the exercise data (e.g., assign the imagery to the variable by clicking the little yellow folder under the "Raster Input File" option). Remember, the exercise imagery is a *.tiff, so you need to change the file type in the Raster Input File navigate to the imagery. Click on "**OK**" (Figure 11.4).

Object Based Image Analysis

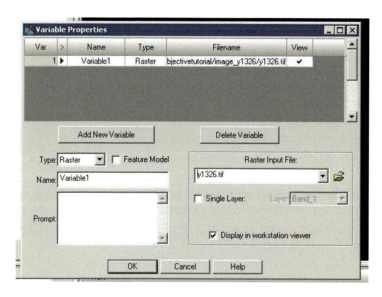

FIGURE 11.4
IMAGINE objective variable properties setup. (From ERDAS IMAGINE®/Hexagon Geospatial.)

4. **Save the project**—In the main IMAGINE Objective interface, click on "**File | Save Feature Project...**" (Figure 11.5).

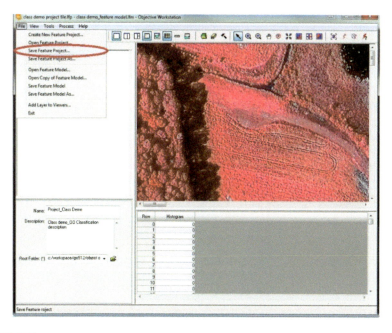

FIGURE 11.5
IMAGINE objective save feature project. (From ERDAS IMAGINE®/Hexagon Geospatial.)

5. Check out the Windows' directory folder where you saved the project and feature model. There should be two new files (Project File (*.lfp) and Feature Model File (*.lfm)) and a new Feature folder. All the subsequent files you create for the classification will be stored in the Feature folder, which also corresponds with the name on top of the Classification Window.
6. Check out the IMAGINE Objective classification interface—There are five main components of the interface:
   a. **Top Toolbar Menu**—Hold the cursor over the tools and see what they are.
   b. **Classification Window**—This is where you define the operators you want to use to do the classification.
   c. **Specification Window**—Here you define metrics for the tools you want to use.
   d. **Image Window**—Here you can define areas of interest, visualize input and output data onto the classification.
   e. **Attribute Table**—Here you see record and field values for pixels and objects (Figure 11.6).

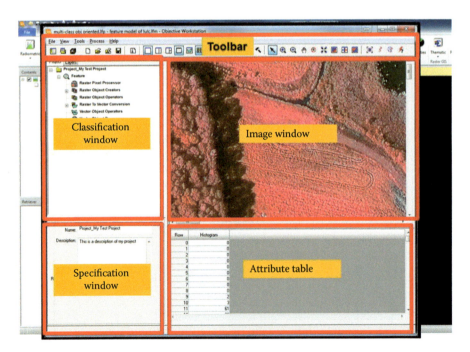

FIGURE 11.6
IMAGINE objective classification interface. (From ERDAS IMAGINE®/Hexagon Geospatial.)

Object Based Image Analysis 237

II. Feature Extraction Procedure

This next set of operations will demonstrate the feature extraction procedure of the classification. Additional information is provided to allow the analyst to make logical sense of each operation. However, this is a **very complex procedure**. So, it is recommended to refer to the help documentation for additional information about other options or choices that may be available.

NOTE: It should also be noted that there are **MANY** different strategies and tools to accomplish the same classification, but for the sake of introducing ourselves to Objective, the methods outlined in this exercise follow a robust set of procedures that should serve well as a base-line for developing a general understanding of the OBIA process. After completing this exercise, go back and try other methods for yourself by consulting the help documentation for guidance.

1. **Classification Window** and **Specification Window**—Click on "feature" in the Classification Window and in the Specification Window check "**Multiple Class.**"

2. Now click on the Classes Tab and define the six classes: (1) **wetlands**, (2) **cultivated**, (3) **forest**, (4) **developed**, (5) **water**, and (6) **barren**. You can type the first-class names into the "**Name**" column by clicking on the default "**Class 1**" in the Class Table. Click on the "+" button to the right of the Class Table to add the second class, and so forth and so on.

3. **Raster Pixel Processor**—Click on "**Raster Pixel Processor**" in the Classification Window and in the Specification Window under "**Available Pixel Cues**" select "**SFP**" (Single Feature Probability).

4. Click on the "+" to add this to the Classification Window. For more information about the Single Feature Probability (SFP) function, you can click on "**?**" to check out the help provided by IMAGINE Objective.

   NOTE: You can also click on "**?**" to find documentation for all other tools as well.

5. Training—Create the training sites. In the Specification Window, click on the **training** tab. An "**Area of Interest (AOI)**" GUI should pop-up. Find the **polygon** button and select the training sites in the image.

6. Add each polygon one at a time and make sure you update the "**Class**" accordingly in the Specification Window (i.e., wetlands, cultivated, forest, etc.) for each polygon you create. Click on the

"**Add**" button at the bottom of the Training Class Table in the Specification Window to add each polygon. When you think you have captured enough representative training samples (at least six sites per class), select "**Accept**."

7. Run the Model—In the Classification Window, "**Right-Click**" on the Raster Pixel Processor. Make sure the "**stop here**" is checked.

You can use "**stop here**" and "**start here**" options to run one process at a time, a couple of processes, or the whole model. This is helpful when you want to make edits to the classification. This exercise will only run one process at a time by specifying "**start here**" and "**stop here**" for each highlighted process in the Classification Window. This is done to ensure that no errors develop in any part of the models along the way, as opposed to entering all the options at one time when selecting to run the entire process in one operation (Figure 11.7).

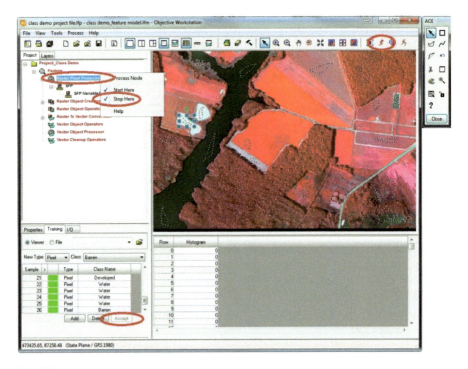

FIGURE 11.7
Designating the "start here" and "stop here" options for running the raster pixel processor model. (From ERDAS IMAGINE®/Hexagon Geospatial.)

Object Based Image Analysis 239

FIGURE 11.8
Output result of the raster pixel processor model. (From ERDAS IMAGINE®/Hexagon Geospatial.)

8. Click on the **lightning bolt** in the upper Tool's menu. This will run the model for this part of the process. You should see an output that looks like Figure 11.8.
9. **Raster Object Creators**—In the Classification Window click on "**Raster Object Creators**."
10. In the Specification Window, add "**Segmentation**" to the Classification Window. Use the default properties. IMAGINE Objective might automatically add segmentation to the Classification Window for a multi-class classification. When you do a single class classification or object extraction there are multiple options. Segmentation clumps similar pixel values in a neighborhood into the same segment or object (some software calls these object primitives).
11. Run the "**Raster Object Creators**," be sure to update start here/stop here. The output should look something like Figure 11.9.

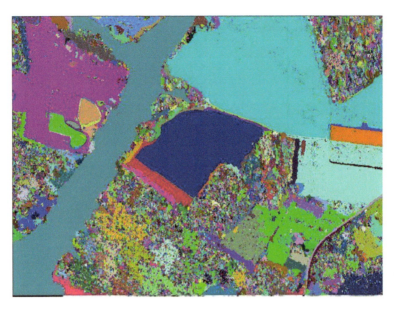

FIGURE 11.9
Output result of the raster object creators model. (From ERDAS IMAGINE®/Hexagon Geospatial.)

Notice the size of the segments. The water body area is quite large as compared to the forest area. This is because the water in this image has a very homogeneous spectral value compared to the forest, which is heterogeneous.

12. Examine the attribute table. You'll notice that each object has a probability of being classified as one of the land cover classes. Use the cursor to click around the segmentation output and the attribute table. Also, take the time to familiarize yourself with some of the data visualization tools on the top toolbar.
13. Find the "**Start Blend**" tool—Over to the right. Click on it. A dialog window like this will pop-up (Figure 11.10).

Object Based Image Analysis

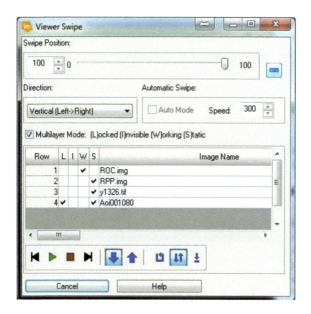

FIGURE 11.10
Viewer swipe tool. (From ERDAS IMAGINE®/Hexagon Geospatial.)

14. Arrange the "[✔]" (check marks) under the **L, I, W,** and/or **S** so that the input imagery appears and disappears below the output of the image segmentation (**ROC.img**).

    You may need to move the AOI layer to the bottom of the available layers in the Classification Window, by clicking the "**Layers**" tab and dragging the AOI layer to the bottom of the list. Review the "**Help**" button for a better understanding of the L (Locked), I (Invisible), W (Working), and S (Static) column headings.

15. You may also want to adjust the "**Blend/Fade**" percentage on the top to swipe and look at the Feature folder. It should be populated with two new outputs: (1) the raster pixel processor (**rpp.img**) and (2) the raster object creator (**roc.img**).

16. **Raster Object Operators**—Click on "**Raster Object Operators**" in the Classification Window. Insert Focal into the Classification Window by highlighting "**Focal**" in the Specification Window and clicking on the "+" to add this function. Try increasing the window size to **9**. Run the "**Raster Object Operators,**" be sure to update start here/stop here. This procedure should remove some of the "**salt and pepper**" artifacts of the output of the image segmentation process.
17. Click on the **Layers** tab in the Classification Window.
18. Right-click on the **ROO_Focal1.img**, click on "**Layer Visibility**"— This will turn off the layer. Notice the difference in the numbers "**salt and pepper**" artifacts in the purple region between the image without the focal procedure (a) and with (b) the focal procedure. You will notice the image on the right has fewer artifacts within the purple region (Figure 11.11).

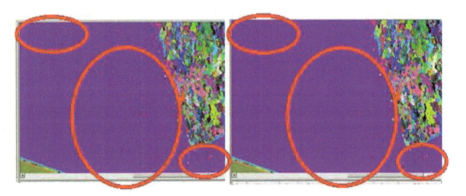

FIGURE 11.11
Examination of results the raster object operators model. Red circles highlight the reduction of artifacts within the purple region. (From ERDAS IMAGINE®/Hexagon Geospatial.)

19. **Raster to Vector Conversion**—Click on "**Raster to Vector Conversion**" in the Classification Window. In the Specification Window add "**Polygon Trace**" to the Classification Window. Run the "**Raster to Vector Conversion,**" be sure to update start here/stop here. This simply creates a vector file out of the input raster file. This process should operate on the last raster you created in the Classification Window (i.e., **ROO_Focal1.img**). However, if you want to perform this operation on a different raster layer available in the Classification Window, click on the "I/O" tab in the Specification Window and adjust the "**Input from Raster Layer**" option. The Raster to Vector Conversion allows you to

Object Based Image Analysis

apply additional shape and pixel-based classification procedures for the wall-to-wall classification. So far, no shape or object-based metrics have been used for deriving the classification. This will occur in the next set of operations. The current output should look like Figure 11.12.

FIGURE 11.12
Output result of the raster to vector conversion model. (From ERDAS IMAGINE®/Hexagon Geospatial.)

20. **Vector Object Operators**—Click on "**Vector Object Operators**" in the Classification Window. From the Specification Window add a probability filter ("+") to the classification tree. Set the Minimum Probability to "**0.00**." Run the "**Vector Object Operators**" and be sure to update start here/stop here. Although this operation is not being invoked, Objective seems to require an input from this process to move onto the next step. Therefore, setting the filter to 0 essentially tells Objective to not to do anything, but it allows the analyst to move onto the next step.

21. **Vector Object Processor**—Click on the "**Available Object Cues**" drop-down box in the Specification Window. There are about two dozen tools here to aid the classification. This is really where the object-based classification begins to take off. For the purposes of this exercise, select "**Geometry:Area**" because it is fairly easy to grasp compared to the other options. Click on the ("+") to add **Geometry:Area** to the classification tree. The next step starts with gathering the area metrics for the Cultivated class.

244   *Image Processing and Data Analysis with ERDAS IMAGINE®*

22. First, click on the "**training**" tab in the Specification Window. Make sure Cultivated is selected as the class. Select a cultivated polygon in the image Window and then click on the "**Add**" button at the bottom of the Specification Window. The idea is to select objects with areas representing the distribution of the object of interest. So the more objects you use in training, the more precise the range of values and classification. Notice that this is now being added as an object, not a pixel. The initial training sites were added as a pixel value. Now the object or shape value can be processed.

23. Select another cultivated polygon and another after that, and so forth for a total of at least three to five polygons.

24. Now click on "**Accept.**"

25. Click on the Distribution tab in the Specification Window. Make sure the class tab is on cultivated. Examine the object count, Minimum, Maximum, Mean, and Standard Deviation. Objective uses this distribution to assign a probability of all polygons fitting the distribution that were just created. If an error occurred here, values can be "hard coded" by simply typing them in.

26. Run the model and be sure to update start here/stop here. Notice the Run Model (red lightning bolt) is grayed out. Click on the "**Show Reasons Why the Feature Model is not Ready to Run**" button in the main process menu. You cannot run the model because of an error. This message is telling you that you need to define object training sites for the five other classes in the classification. So now repeat this process for other classes.

27. Accept the object training sites and run the model, be sure to update start here/stop here. The important output is the attribute table. Take a look. Now you have an object probability and a pixel probability for each object.

28. **Vector Cleanup Operators**—Click on "**Vector Cleanup Operators**" in the Classification Window. Add "**Label**" ("+") in the Specification Window and run the model, be sure to update start here/stop here. This is the final wall-to-wall classification. It classifies each object based on the probability that the object is one of the six objects being classified—the class that has the highest probability is simply used to label (or classify) that object.

29. Once you have saved the data, you may close IMAGINE Objective.

Object Based Image Analysis 245

Examine the Object-Oriented Classification in ERDAS IMAGINE

1. In ERDAS IMAGINE's Table of Contents, open the "**VCO_Label1.shp**" dataset (remember this is vector layer; **File | Open | Vector Layer**). Examine the attribute table to see the objects classified. Click around on the objects to see what you got right and what you got wrong.

   The following image details what a few of the selected objects from the attribute table looked like water (a) and cultivated (b) when processing the example data (Figure 11.13).

FIGURE 11.13
Selected objects (yellow) from the attribute table. The selected attribute in the image (left) is water and cultivated in right image. (From ERDAS IMAGINE®/Hexagon Geospatial.)

2. Assign a classification color scheme to the new object-oriented classification. Expand the "+" symbol next to the "**VCO_Label1.shp**" dataset in the ERDAS IMAGINE's Table of Contents. To change each of the "**Filled Solid Color**" class categories from white to the representative color scheme as follows, highlight the first "**Filled Solid Color**" class category in the Table of Contents and select a representative color from the available color options in the Style menu ribbon within the Vector Tab grouping.

3. Repeat this process for each of the five remaining class categories. The following image represents a completed IMAGINE Objective classification (Figure 11.14).

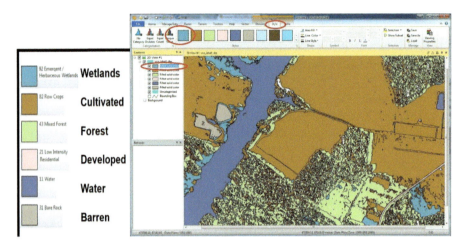

FIGURE 11.14
Vector style options used to assign the classification color scheme. (From ERDAS IMAGINE®/Hexagon Geospatial.)

The results of this OBIA classification resulted in six land use/land cover classification types (Water, Developed, Barren Land, Forest, Cultivated Land, and Wetlands). The water land cover class represented the most homogeneous areas within the image when examining the results of the Raster Object Creators—Segmentation function. All the land cover classes required use of the Raster Object Operators—Focal function to reduce some of the *salt and pepper* artifacts in the output of the image segmentation process. Whereas, the cultivated and forest land cover classes also resulted in reduced, but not completely eliminated artifacts using the specified window size setting of 9. Increasing this setting to a slightly higher value, such as 10 or 11, may provide an improved reduction of artifacts within these class categories, but at a cost of losing finer classification details, primarily along the edges of each category. However, with the result of any classification, the classification output results should be evaluated based on the intended purpose of the final classification. Also an accuracy assessment should be performed to gauge whether the classification output results fit the intended purpose.

Review Questions

1. Why do object-oriented classification, image extraction, or Object Based Image Analyses (OBIA), in certain cases, seem to be more useful for higher-spatial resolution imagery than the traditional pixel-based approach provided by the supervised or unsupervised classification methods?

Object Based Image Analysis 247

2. In general terms, how does the OBIA methods differ from the traditional pixel-based approach provided by the supervised or unsupervised classification methods?
3. What are the image-based *cues* that are used an OBIA approach?
4. What are classification-based objects that are used an OBIA approach?
5. Name the statistical algorithms incorporated in an OBIA approach to complete the classification operation after each object has been identified?

12

Additional Image Analysis Techniques

Overview

In the process of generating an effective land use/land cover classification, additional analysis techniques are often required to extract as much information as possible from the imagery data. There are a number processes that may involve several individual or combined techniques in order to achieve your classification goals. These techniques may include generating a water-only image mask, which is useful for masking out, or removing, the water from an image. A water-only image mask may be particularly useful in an image where water pixel values encompass a high spectral range. Another useful technique is creating a Normalized Difference Vegetation Index (NDVI), or greenness map.

The NDVI it is a popular tool used in forest vegetation health assessments and represents an index of vegetation greenness. Chlorophyll pigments (light absorbing pigments) found in the leaves of healthy vegetation absorb incoming radiation (visible light) in the blue (0.45 μm) and red range (0.67 μm) of the electromagnetic spectrum (EMS). Green (0.5 μm) is reflected more so in the near infrared (0.7–1.3 μm) ranges of the EMS. Particularly in the near infrared portion of the EMS, there is a strong reflection of light largely due to the internal makeup of a leaf's structure. When vegetation becomes stressed, due to drought, infestation, disease, and other reasons, the leaves typically decrease reflection in the near infrared range as the leaf's internal structure begins to change. A NDVI image is created for specific image dates through the division of the image's red and near infrared spectral bands ($NIR - RED/NIR + RED$). Multiple NDVI images can be generated to determine the spectral signatures for areas of healthy vegetation vs. areas of stressed vegetation. This type of application is useful for identifying areas where vegetation is under stress.

249

250   *Image Processing and Data Analysis with ERDAS IMAGINE®*

Additional Analysis Techniques Application

This exercise will complete several unsupervised classifications of San Francisco, California using the Landsat 8 subset image created in the previous chapter. This exercise will involve creating image subsets, recoding, binary image masking, and using other ERDAS modeling tools, such as the Normalized Difference Vegetation Index or (NDVI).

Learning Objectives

1. Explore additional classification techniques.
2. Create subsets, recoding, masking and using other ERDAS modeling tools.

Data Required: Landsat 8 image **LC08_L1TP_044034_20170716_20170727_01_T1**, Acquisition Date: 2017/07/16, of San Francisco, California. This image was created from raw data that was downloaded from the United States Geological Survey (USGS) EarthExplorer (http://earthexplorer.usgs.gov/) website following the procedures demonstrated in the previous chapter. Additionally, Chapter 1 provides greater detail on searching and acquiring data through the EarthExplorer website.

Once the San Francisco Landsat 8 image has been downloaded, create a multi-band image following the procedures described in the previous chapter. Next, subset the image to identify a region in the upper northwestern quadrant of the San Francisco Peninsula as seen in the following image (Figure 12.1).

Additional Image Analysis Techniques

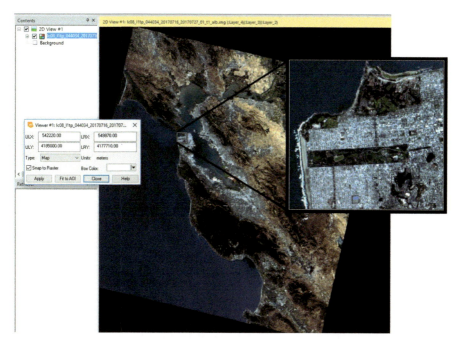

FIGURE 12.1
Image subset region for the Landsat 8 image (LC08_L1TP_044034_20170716_20170727_01_T1; Acquisition Date: 2017/07/16) of San Francisco, California. (From ERDAS IMAGINE®/Hexagon Geospatial.)

I. Create a Land-Only Image

Next, a land-only image will be created in three procedural operations to remove water from the image. Often the presence of water within an image can make land cover classification difficult, as water may sometimes encompass a very wide spectral range. By removing water and creating a land-only image, the statistical operations of the classification procedure can focus only on the terrestrial values of the remaining land pixels. The three procedural operations used to create the land-only image include, an unsupervised classification, a thematic image recode, and a binary image mask to remove the water from original Landsat 8 image subset.

Initiating the Classification Procedure

1. If not already open, open ERDAS Imagine. Use **File | Open | Raster Layer** to display the San Francisco, Landsat 8 subset image as a False Color Composite (FCC) (5, 4, 3).

2. Create a new "**unsupervised**" classification of the Landsat 8 subset image, as follows:

   - Select **Raster | Unsupervised | Unsupervised Classification**.

   - Specify an output file name of **sanfrancisco_unsupclass** in **Output Cluster Layer** in a folder you can find easily (preferably where the initial images have been saved). It is not necessary to save the output signature file so do not check Output Signature Set.

   - For this exercise, only **100** output classes will be specified to demonstrate the issue of spectral clustering that may occur if too few classes are chosen. As mentioned before in the Unsupervised Classification exercise, ordinarily a good rule of thumb is to start with roughly 200 output classes for each intended category you hope to achieve in the final unsupervised thematic map.

   NOTE: Even 200 output classes per class may be a modest choice. Additionally, it is a good idea to use the Spectral Profile tool (Chapter 6) to examine representative classes throughout the entire raw image to get an idea of how similar or variable the spectral values may be for each class. However, for this example, use 100 output classes.

   - Click on the **Color Scheme Options** button and change the output color values to 5, 4, 3 to match the Landsat 8 subset image's FCC display (Figure 12.2).

   - Keep all the rest of the settings as default. Then click **OK**.

   - A Process List window should appear and begin the procedure. Once processing is complete, close the process list window.

Open the unsupervised thematic/classified image (**File | Open | Raster Layer...**) and select the **sanfrancisco_unsupclass** image.

NOTE: If you accepted the default "**Color Scheme Options...Approximate True Color**" during the Unsupervised Classification, the output color coding of the classified image follows the color scheme of the Landsat 8 FCC subset image. Therefore, the unsupervised classified **sanfrancisco_unsupclass** image will look almost similar to the original image.

Additional Image Analysis Techniques

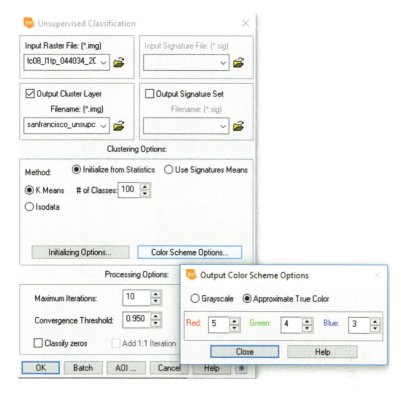

FIGURE 12.2
Unsupervised classification options. (From ERDAS IMAGINE®/Hexagon Geospatial.)

Create a Thematic Image Recode

Next, **recode** the values assigned to each class. This exercise will recode the initial 100 classes into just two classes, either **land** = 2 or **water** = 1.

1. To explore which classes are water and which are land, right-click on the **sanfrancisco_unsupclass** image and display the attribute table. Change the color of each class by clicking on each row in the attribute table to see which classes are "**water**" and which are "**land**."
2. In the "**Class_Name**" column of the attribute table, change all classes you determine to represent water to "**1**" and all classes you determine to represent land to "**2**." Be sure to save the work once complete by **right-clicking on the file name** in the Contents Legend and then **Save Layer**. Next, close the attribute table.

    As you can see from the following image, the following seven rows were identified as water: 1, 2, 3, 9, 12, 37, and 40. The results may vary depending on the amount of area included within the subset image. Do not change the box labeled "**0**" or "**unclassified**" (Figure 12.3).

254 *Image Processing and Data Analysis with ERDAS IMAGINE®*

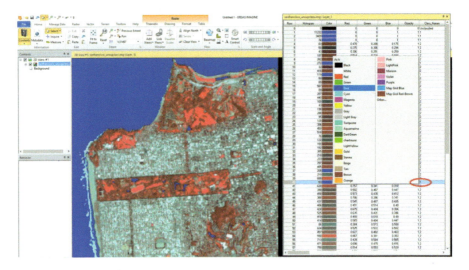

FIGURE 12.3
Class and color assignments for unsupervised classification output results. (From ERDAS IMAGINE®/Hexagon Geospatial.)

> **NOTE:** Some of the urban areas may have been *grabbed*, or misclassified, by the water class. The only way to reduce this error is to increase the **number of "classification clusters."**

3. Click on the **Raster | Thematic** within the Raster GIS category grouping | **Recode** button. Make sure that **sanfrancisco_unsupclass** is the input file. Name the output file **sanfrancisco_recode**. Make sure that **"Ignore Zero in Stats"** is selected. Do not close the Recode dialog window yet (Figure 12.4).

FIGURE 12.4
Thematic recode options. (From ERDAS IMAGINE®/Hexagon Geospatial.)

Additional Image Analysis Techniques 255

4. Click on the "**Setup Recode**" button. In order to recode the image to only differentiate water from land, select and highlight the water classes (use **Shift-Click** to select multiple classes). Next change the number in the "**New Value**" column for these rows to "**1**" by clicking on "**Change Selected Rows**."

5. Next, highlight the land classes and change the number in the "**New Value**" column for these rows to "**2**." Click on **Change Selected Rows** and click **OK**, and **OK** again in the Recode dialog window. Now the new classes will simply be displayed as water = 1 or land = 2 (Figure 12.5).

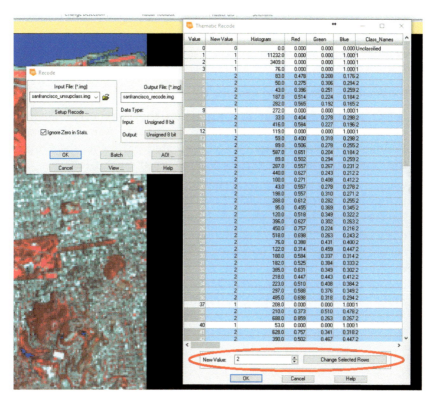

FIGURE 12.5
Thematic recode change selected rows options. (From ERDAS IMAGINE®/Hexagon Geospatial.)

6. Open the newly recoded classified image (**File | Open | Raster Layer**) and select the **sanfrancisco_recode** image. Open the attribute table and change the color of class 1 to blue and class 2 to tan to view the two new classes. You may also wish to inspect the image by comparing it to the original subset image using the **Swipe** utility tool (Figure 12.6).

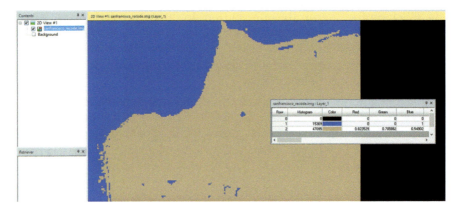

FIGURE 12.6
Thematic recode results. Class 1 (blue) represents water waters and class 2 (tan) represents land features. (From ERDAS IMAGINE®/Hexagon Geospatial.)

Create a Binary Image Mask to Remove Water

1. Select **Raster | Subset & Chip** from the Geometry category grouping **| Mask**. Select the original **San Francisco, Landsat 8 subset image** that you created earlier as the "**input file**" and **sanfrancisco_recode** as the "**input mask file**."
2. Click on "**setup Recode**." Next click on class "**1**" in the **Value** column to highlight it (remember class 1 was the water features). Enter "**0**" in the **New Value** box and click on "**Change Selected Rows**."
3. Click on class "**2**" in the **Value** column to highlight it. Then change class 2 to a value of "**1**" (remember class 2 was the land features) in the **New Value** box and click on "**Change Selected Rows**." Name the output file **sanfrancisco_land**. Make sure "**Ignore Zero in Output Stats**" is selected and click **OK** (Figure 12.7).

Additional Image Analysis Techniques

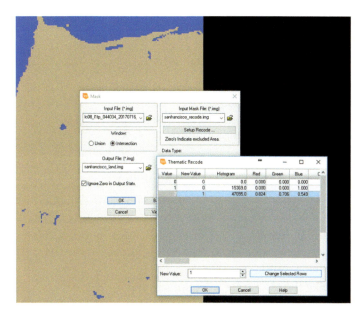

FIGURE 12.7
Thematic recode setup to remove or *mask* water (water recoded as 0). (From ERDAS IMAGINE®/Hexagon Geospatial.)

4. Next, add the new **sanfrancisco_land** file as a FCC (5, 4, 3) to examine it (swiping between sanfrancisco_recode and sanfrancisco_land files may be helpful). Now you have created a classification mask image that contains only the land features, as the water features are now *masked out* (Figure 12.8).

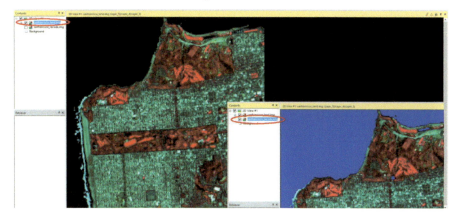

FIGURE 12.8
Results of water image mask. Upper display shows sanfrancisco_land image as a FCC (5, 4, 3) with water removed. Lower image (inset) shows the land and water thematic recoded image (sanfrancisco_recode) below the water mask image (sanfrancisco_land). (From ERDAS IMAGINE®/Hexagon Geospatial.)

II. Create a Normalized Difference Vegetation Index

Now that the water has been removed from the image, a specialized type of unsupervised classification will be performed on just the land area. This specialized type of unsupervised classification is a vegetation index representing the amount of greenness within the image and is known as a Normalized Difference Vegetation Index or *NDVI*.

Create a Normalized Difference Vegetation Index

1. Clear the display and click on **Raster | Unsupervised** within the Classification category grouping | **NDVI**. Use **sanfrancisco_land** as the input file (remember, this file contains only the land features as water was masked out in the previous steps). Name the output file **sanfrancisco_ndvi**.

2. Next, make the following selections in the Indices dialog window:

   Senor: select **WorldView-2 Multispectral**

   NOTE: The image that the **sanfrancisco_land** image was originally derived from came from the Landsat 8 sensor. Because this option does not show up, it is necessary to ensure the correct bands that relate to the Landsat 8 (Red (Band 4) = 636–673 nanometers and NIR (Band 5) = 851–879 nanometers) are specified.

   Category: Vegetation

3. **Index:** NDVI—Normalized Difference Vegetation Index as the function, thus the formula should be: **(NIR − RED)/(NIR + RED)** (Figure 12.9).

   Band Selection: In the NIR column, check Band 7 and in the Red column check Band 5.

Additional Image Analysis Techniques

FIGURE 12.9
Indices options. (From ERDAS IMAGINE®/Hexagon Geospatial.)

4. In **I/O Options** tab, select **Map** as the coordinate type and check **Stretch to Unsigned 8 bit**. Leave all other options at the default settings. Then select **OK** to begin the NDVI classification (Figure 12.10).

FIGURE 12.10
NDVI classification setup. (From ERDAS IMAGINE®/Hexagon Geospatial.)

5. Next, add the new **sanfrancisco_ndvi** raster to the viewer (Figure 12.11).

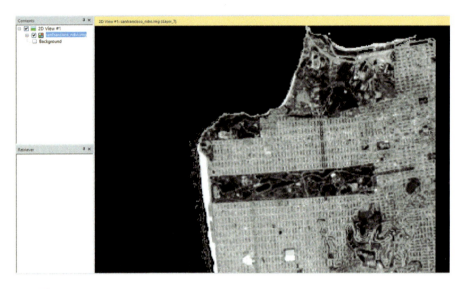

FIGURE 12.11
Output result of the NDVI classification. (From ERDAS IMAGINE®/Hexagon Geospatial.)

Classify the Normalized Difference Vegetation Index

6. Next, perform an unsupervised classification on the previously created **sanfrancisco_ndvi**.

Select **Raster | Unsupervised** within the Classification category grouping | **Unsupervised Classification**. Specify an output file name of **sanfrancisco_veg**. Specify the number of output classes. In this case, only do **10**. Next, select **Color Scheme Options** and choose **Grayscale**.

An output signature file is not necessary, so uncheck the box for this option. Then click **OK** (Figure 12.12).

Additional Image Analysis Techniques   261

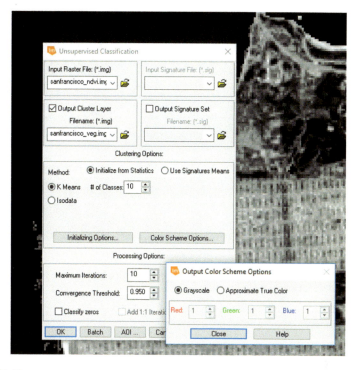

FIGURE 12.12
Unsupervised classification of the sanfrancisco_ndvi output layer. (From ERDAS IMAGINE®/Hexagon Geospatial.)

7. A Process List window should appear and begin the procedure. Once processing is complete, **close** the process list window.
8. Open the newly classified image (**File | Open | Raster Layer**) and select the **sanfrancisco_veg** image.
9. Open the attribute table for **sanfrancisco_veg** by right-clicking on the raster.

    You should see 10 classes and one "**unclassified**" labeled as **0**.

10. Change the colors of classes 1–10 and try to differentiate between different vegetation classes. You should be able to separate a minimum of five classes (described in the previous chapter): **deciduous forest**, **evergreen forest**, **herbaceous grass**, **developed (non-vegetated)**, and **barren land (mostly beech areas)**.
11. Recode if necessary to combine classes if one of the vegetation groups is spread across multiple classes (Figure 12.13).

FIGURE 12.13
Output result of the unsupervised classification of the sanfrancisco_ndvi output layer. (From ERDAS IMAGINE®/Hexagon Geospatial.)

III. Create an Impervious Surface Map

Following the completion of the NDVI output, next explore the data a little further to determine if the developed areas (such as impervious land cover) can be can better isolated by removing vegetation!

Create an Impervious Surface Map (Remove Vegetation)

1. Next, all vegetative communities will be masked out leaving only the *developed* features remaining in the image.
2. Select **Raster | Subset & Chip | Mask**. Select the San Francisco, Landsat 8 subset image as the input file and **sanfrancisco_veg** as the input mask file.
3. Click on **setup Recode** and highlight all the vegetative classes (such as deciduous and evergreen). Enter "**0**" in the New Value box and click on "**Change Selected Rows.**" Next, highlight all the developed classes (such as developed and barren) and enter "**1**" in the New Value box and click on "**Change Selected Rows**" (Figure 12.14).

Additional Image Analysis Techniques

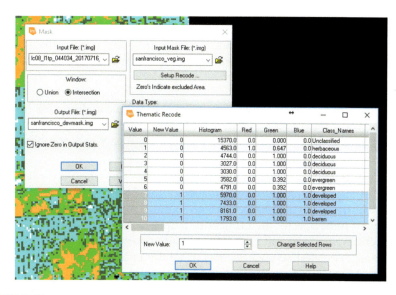

FIGURE 12.14
Thematic recode to create a vegetation mask (remove impervious surface). All vegetation class values are recoded to 0. (From ERDAS IMAGINE®/Hexagon Geospatial.)

4. Name the output file **sanfrancisco_devmask**. Make sure **Ignore Zero in Output Stats** is selected. Select OK to create the mask. Next, examine the new **sanfrancisco_devmask** by adding it to the display (swiping between sanfrancisco_devmask and sanfrancisco_veg files may be helpful) (Figure 12.15).

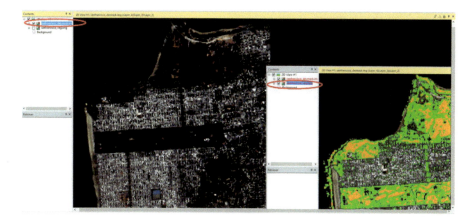

FIGURE 12.15
Results of vegetation image mask. Upper display shows the sanfrancisco_devmask layer. Lower image (inset) shows the sanfrancisco_veg layer below the sanfrancisco_devmask layer. (From ERDAS IMAGINE®/Hexagon Geospatial.)

Classify the Impervious Surface Map

Next, an unsupervised classification will be performed on just the *developed*, or non-vegetated, categories using an unsupervised classification and the mask created in the previous step.

5. Decide on a band combination for the **sanfrancisco_devmask** image that you think would separate different intensities of development or types of impervious surface (i.e., 5, 4, 2). Apply this combination to the image in the 2D View window.

6. Select **Raster | Unsupervised** within the Classification category grouping | **Unsupervised Classification**. Specify the input file as **sanfrancisco_devmask** and an output file name **sanfrancisco_impervious**. Specify the number of output classes. In this case, choose 6 for low, medium and high-intensity urban cover. An output signature file is not needed, so uncheck the box for this option if it is checked. Also adjust **the Color Scheme Options | Approximate True Color** to match the band combinations (e.g., 5, 4, 2). Then click **OK** (Figure 12.16).

7. A Process List window should appear and begin the procedure. Once processing is complete, close the process list window. Open the newly classified image (**File | Open | Raster Layer**) and select the **sanfrancisco_impervious** image (Figure 12.17).

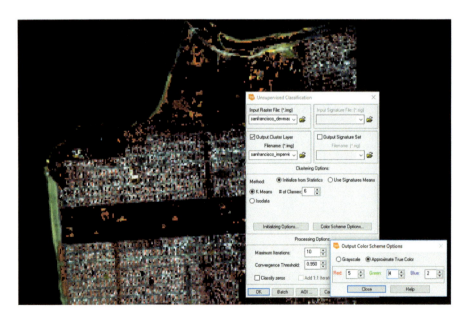

FIGURE 12.16
Unsupervised classification of the sanfrancisco_devmask output layer. (From ERDAS IMAGINE®/Hexagon Geospatial.)

Additional Image Analysis Techniques 265

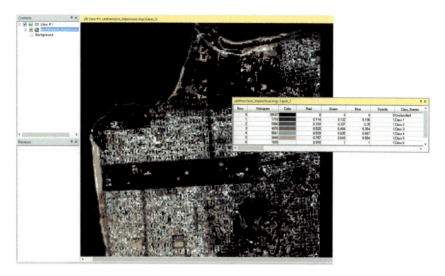

FIGURE 12.17
Output result of the sanfrancisco_impervious unsupervised classification. (From ERDAS IMAGINE®/Hexagon Geospatial.)

8. Zoom into the image to view the different class of impervious cover. You can change each classes color in the attribute table if needed (Figure 12.18).

FIGURE 12.18
Modified class color values of the sanfrancisco_impervious unsupervised classification. (From ERDAS IMAGINE®/Hexagon Geospatial.)

Recombine Classification Components

1. You may choose to recombine the three classification components in the viewer (water-only, NDVI-veg and other-Impervious) for a final output or more detailed classification.

2. Try by first visualizing a combined dataset by selecting **sanfrancisco_unsupclass**, **sanfrancisco_veg**, and **sanfrancisco_impervious** in the 2D View window (Figure 12.19).

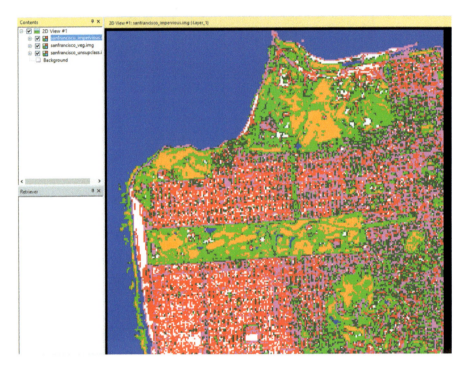

FIGURE 12.19
Display of the three classification layers (sanfrancisco_impervious, sanfrancisco_veg, and sanfrancisco_unsupclass) in the 2D View display. (From ERDAS IMAGINE®/Hexagon Geospatial.)

3. You may even want to go back to the original unsupervised classification and mask out the land. Try seeing if you can separate water from dark, wet soil.

Combining Output Layers in Model Maker

It is also possible to explore options with recoding each output layer and then combining them in the model maker. For example, if the analyst wants to possibly create an image to (perhaps) strongly accentuate impervious features, the

Additional Image Analysis Techniques 267

sanfrancisco_impervious image (containing only impervious features masking out all other non-impervious features) could be added to the recoded San Francisco image (the **sanfrancisco_veg** file containing all 10 land cover class, plus the unclassified class). The resultant image would produce a combined image that has the impervious-only layer added into the vegetation recoded original image, which also contains the impervious data, thus *theoretically* increasing/improving the impervious signature (Figure 12.20).

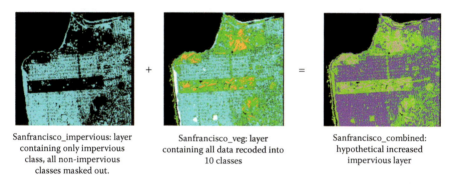

Sanfrancisco_impervious: layer containing only impervious class, all non-impervious classes masked out.

Sanfrancisco_veg: layer containing all data recoded into 10 classes

Sanfrancisco_combined: hypothetical increased impervious layer

FIGURE 12.20
Recombination of classification layers sanfrancisco_impervious and sanfrancisco_veg, to the output layer, sanfrancisco_combined). (From ERDAS IMAGINE®/Hexagon Geospatial.)

Here are the steps for performing this operation with Model Maker:

1. First start Model Maker: **Toolbox** tab | **Model Maker** (Figure 12.21).

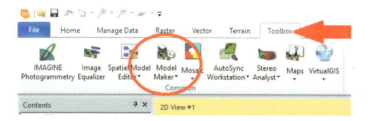

FIGURE 12.21
Model Maker toolbox option. (From ERDAS IMAGINE®/Hexagon Geospatial.)

2. Next, in the "New Model" dialog box, select the "**Place a Raster object in the model**" object in the model space. Do this for the two input image layers and your output image layer (Figure 12.22).

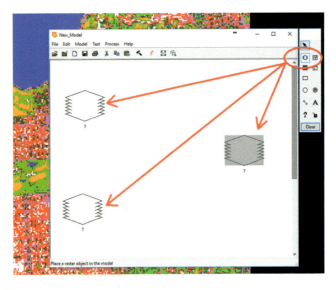

FIGURE 12.22
Model maker raster layer options within New Model window. (From ERDAS IMAGINE®/Hexagon Geospatial.)

3. Double-click on each Raster Object to specify the input layers (**sanfrancisco_impervious** and **sanfrancisco_veg**), as well as name the output image layer (such as **sanfran_combined_veg_imperv**) (Figure 12.23).

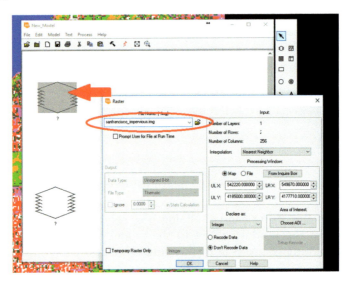

FIGURE 12.23
Assign raster input and output layers to New Model window. (From ERDAS IMAGINE®/Hexagon Geospatial.)

Additional Image Analysis Techniques 269

4. Next, select the "**Place a function in the model**" ◯ option, and use the cursor to draw a function circle in the middle of the model space as seen in the following diagram. Also, select the "**Connect inputs to function or function to outputs**" option, and use the arrow in the model space to connect the two input image layers (**sanfrancisco_impervious** and **sanfrancisco_veg**) to the function and a single arrow to connect the function to the output raster object (**sanfran_combined_veg_imperv**) (Figure 12.24).

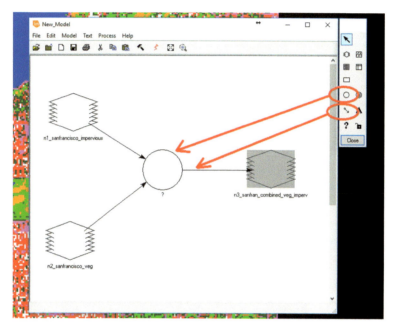

FIGURE 12.24
The red arrows show options to insert the "Place a function in the model" option and "Connect inputs to function or function to outputs options." (From ERDAS IMAGINE®/Hexagon Geospatial.)

5. Next, double-click the function circle in the model space to open the "**Function Definition**" dialog box. In the **Function Definition** dialog box, select "**Arithmetic**" from the "**Function**" drop-down list and create the expression: **$n2_sanfrancisco_veg + $n1_sanfrancisco_impervious** (Figure 12.25).
6. Click OK in the "Function Definition" dialog and the red lightning bolt in the "New Model" window to run the model.

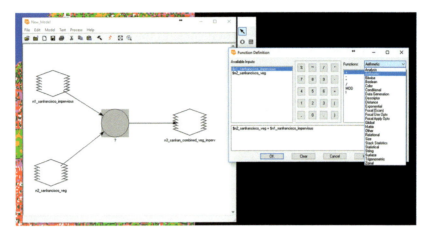

FIGURE 12.25
Function Definition dialog box is used to create the arithmetic expression "$n2_sanfrancisco_veg + $n1_sanfrancisco_impervious," which will recombine the two input layers. (From ERDAS IMAGINE®/Hexagon Geospatial.)

7. Once the "Process List" is done, close this list and the "New Model" window. Then open the new image layer in ERDAS Image. You may compare this image to the original **sanfrancisco_impervious** image (Figure 12.26).

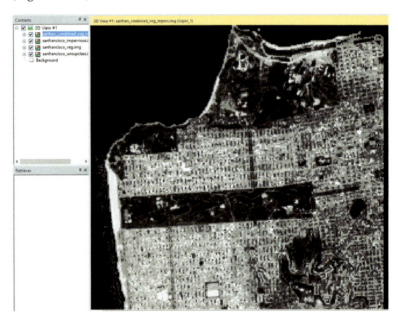

FIGURE 12.26
Output result of the recombined layer sanfran_combined_veg_imperv. (From ERDAS IMAGINE®/Hexagon Geospatial.)

Additional Image Analysis Techniques 271

Review Questions

1. Often additional analysis techniques are required to extract as much information as possible from the imagery data. There are a number processes that may involve a number of the individual or combined techniques in order to achieve your classification goals. Name a couple of additional analysis techniques discussed in this chapter that may be useful in the process of generating an effective land use land cover classification.

2. What is the water-only image mask useful for and how is it constructed?

3. What is the Normalized Difference Vegetation Index (NDVI) useful for?

4. Discuss how the NDVI operation works.

5. How is a NDVI useful for identifying areas where vegetation may be under stress?

13

Assessing Thematic Classification Accuracy

Overview

An important component of any land use/land cover classification is the assessing of the accuracy of the classified data. It is important to note that results of a classification or output map represents an imperfect depiction of the original data. All classification outputs contain errors, and it is the responsibility of the remote sensing analyst to characterize these errors prior to a map's use in subsequent applications. The most widely accepted method for the accuracy assessment of remote-sensing-derived maps is by comparison to reference data (also known as *ground truth*) collected by visiting an adequate number of sample sites in the field (Goodchild et al. 1992, Congalton and Green 1999, Khorram et al. 1999). The key instrument in this comparison is the generation of an accuracy assessment. This assessment includes an error matrix, which quantifies the accuracy for each map class of interest as well as the overall map accuracy (such as combining all the classes) and the concepts of Producer's Accuracy, User's Accuracy, and the Kappa statistics (Khorram et al. 2016b). The accuracy assessment represents the reliability and theoretical repeatability of the final thematic classification generated by your methods. In a land cover classification project, an error matrix is typically produced to demonstrate class accuracies based on your classified categories. Diagonal elements of the error matrices are the number of sites correctly classified in the image. The sum of off-diagonal elements in each column represent the sites not identified as belonging to a particular class. This type of error is known as an *omission error*. The sum of the off-diagonal elements in each row represent the sites belonging to a particular class when they actually belonged to a different class. This type of error is known as a commission error. Error matrices are useful for determining classification categories, or classes, that are most likely to be misclassified, or confused, and are sometimes referred to as Confusion Tables.

The error matrix produces two types of accuracy estimates. The first of these is referred to as **"Producer's Accuracy."** The Producer's Accuracy is the probability that an area that is in class "X" has been correctly identified as being in class "X". This accuracy represents possible errors of omission as it defines the number of verification sites identified in the classification. The second accuracy type is known as the **"User's Accuracy."** The User's Accuracy is the probability that an area that has been classified as "X" actually is in class "X". These types of errors represent errors of commission as it defines the number of verification sites *committed* to the correct class. Both Producer's and User's Accuracy estimates are useful depending on the intended use of the data. As well, these accuracy estimates can be an indication of possible limitations of the data itself.

Standard errors for categories represent the amount of variation associated with each estimate of a class accuracy. The standard error values can be used to determine a confidence interval for an estimated class accuracy. For example, 95 times out of 100, water will be classified with a User's Accuracy of 94.89%–99.11% (97% plus or minus 2.11). When the number of sample sites is low, the degree of confidence in an estimate tends to decrease (e.g., the standard error tends to increase, and the confidence interval widens). A class accuracy cannot be greater than 100%. However, when no errors are found during the classification procedure, no estimate of error can be determined.

I. Assessing Thematic Classification Accuracy Application

Learning Objectives

1. To demonstrate procedures for determining the thematic accuracy of image interpretation.
2. To become more familiar with the terminology used in accuracy assessment.

Data Required:

- The thematic supervised classification image, created in Chapter 10 from the Landsat 8 subset image of the original United States Geological Survey (USGS) EarthExplorer downloaded file (**LC08_L1TP_191030_20170706_20170716_01_T1**, Acquisition Date: 2017/07/06), for the town Orvieto, in the Province of Terni, Italy.

Assessing Thematic Classification Accuracy 275

- The Sentinel-2 subset image, created in Chapter 10, of the original USGS EarthExplorer downloaded file (**L1C_T32TQN_A010601_20170703T101041**, Acquisition Date: 2017/07/03) for the town Orvieto, in the Province of Terni, Italy.

II. Recoding the Supervised Classification

If the data is not already recoded (previously described in Chapter 12), it may be easier to convert the thematic supervised classification image into a "**RECODED**" classified dataset. The recode operation will reduce the multiple supervised training sites (the roughly 50–60 Area of Interest (AOI)/polygons per class of homogenous pixels) to the number of category classes that represent the least number of representative classes (such as six land cover classes). For example, if you have 50–60 training samples in the classified image for water, 50–60 for developed, 50–60 for deciduous, and so forth, you may want to use the **Thematic Recode** option to group, or recode, the supervised classified image to six land cover category classes.

1. In the 2D View window, open the thematic supervised classification image, created in Chapter 10 from the Landsat 8 subset image.
2. Next, select the **Raster** tab | **Thematic** within the Raster Geographic Information System (GIS) grouping | **Recode**. Make sure the Landsat 8 subset image thematic image is the input file. Name the output file *Orvieto_alb_subset_supclass_recode* (or something similar). Make sure that "**Ignore Zero in Stats**" is selected. Do not close the Recode dialog window yet.
3. Click on the "**Setup Recode**" button. Select and highlight the water classes (use **Shift-Click** to select multiple classes). Next change the number in the "**New Value**" column for these rows to "**1**" by clicking on "**Change Selected Rows**".
4. Next, highlight the remaining classes and change the numbers in the "**New Value**" column for these rows as follows to recode the supervised classified image into six classes: 2—developed, 3—evergreen, 4—deciduous, 5—agriculture/herbaceous, 6—barren. Click on **Change Selected Rows** for each New Value and click **OK**, and **OK** again in the Recode dialog window (Figure 13.1).

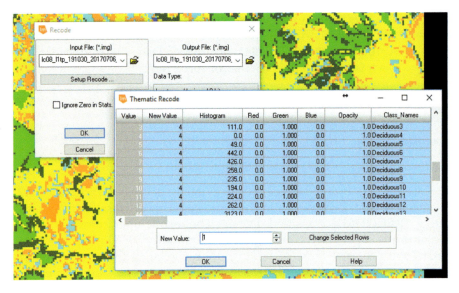

FIGURE 13.1
Thematic recode, into 6 land cover category classes, of the original Landsat 8 OLI/TIRS subset image (LC08_L1TP_191030_20170706_20170716_01_T1) for the town Orvieto, in the Province of Terni, Italy. (From ERDAS IMAGINE®/Hexagon Geospatial.)

5. Display the newly created recoded image file. Now open the attribute table (**Right-Click** on the file name in the contents Legend | **Display Attribute Table**) and change the row colors within the attribute table as follows:

Row 0: Unclassified—Black

Row 1: Water—Blue

Row 2: Urban/Developed Land—Cyan

Row 3: Evergreen Vegetation—Dark Green

Row 4: Deciduous Vegetation—Green

Row 5: Herbaceous Vegetation—Yellow

Row 6: Barren Land—Orange (Figure 13.2)

Assessing Thematic Classification Accuracy 277

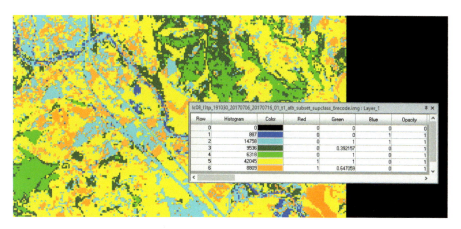

FIGURE 13.2
Output result of the thematic recode of the 6 land cover category classes. (From ERDAS IMAGINE®/Hexagon Geospatial.)

6. Save the changes (**Right-Click** on the file name in the contents Legend | **Save layer As**).

III. The Accuracy Assessment Procedure

1. Open the Sentinel-2 subset image, created in Chapter 10, of the original EarthExplorer downloaded file (L1C_T32TQN_A010601_20170703T101041, Acquisition Date: 2017/07/03) for the town Orvieto, in the Province of Terni, Italy (**Orvieto_b2-4_8_subset.img**). The unclassified or raw, Sentinel-2 subset image will be used as a *"true reference"*, meaning that locations/land covers that you interpret from this image are considered accurate. Ideally, if available, you would want to use a separate data source as a reference (e.g., high-resolution air photo, GPS coordinates, field visits). However, the Sentinel-2 dataset, being acquired at a higher spatial resolution (10 m) than the Landsat 8 image (30 m) that was used for the supervised classification will serve as a point of demonstration. Additionally, the Sentinel-2 reference image may be displayed as True Color Composite (TCC) band combination (3, 2, 1), or as False Color Composite (FCC) band combination (4, 3, 2) to improve the analysts' ability to interpret the *true classes* on the ground within the reference image.

278   *Image Processing and Data Analysis with ERDAS IMAGINE®*

2. Also, in the 2D View #1 viewer, open the recoded Orvieto thematic supervised classification image that was created earlier in Chapter 10 (i.e. *Orvieto_alb_subset_supclass_recode*, or whatever naming convention you used for the completed Orvieto thematic supervised classification image). The recoded classified image should display on top of the Orvieto_b2-4_8_subset.img in the viewer.

3. Next, open the Accuracy Assessment tool: **Raster** tab | **Supervised** within the Classification category grouping | **Accuracy Assessment**. In the Accuracy Assessment tool window that opens, open the Orvieto_alb_subset_supclass_recode (File | **Open** | *Orvieto_alb_subset_supclass_recode*).

4. Select the following options within the Accuracy Assessment tool window:

   - **View | Select Viewer...** (click anywhere in the viewer). This procedure selects the viewer to display random points.
   - **Edit | Create/Add Random Points**. The procedure adds random points to the image within the selected display.
   - Within the Add Random Points window, select "**Stratified Random**" within the Distrusted Parameters grouping. This ensures points are randomly distributed among selected classes.

   In the Number of Points box, enter **300** since six land cover categories were classified.

   NOTE: A good rule of thumb for a classification with a total overall accuracy of 85%: at least **250 points or more**, with a minimum of **50 points per class** to ensure each class is within 5% of the total estimated classification error should be used. The 250 points requirement is only the recommended minimum number of points that should be used for any accuracy assessment if you are hoping to achieve the total overall accuracy of 85% or better. In addition, you should also ensure you have 50 points per class to reach this accuracy goal within the intended 5% range for each class. Depending on the number of classes, you may need **a lot** more than 250 points (Figure 13.3).

Assessing Thematic Classification Accuracy 279

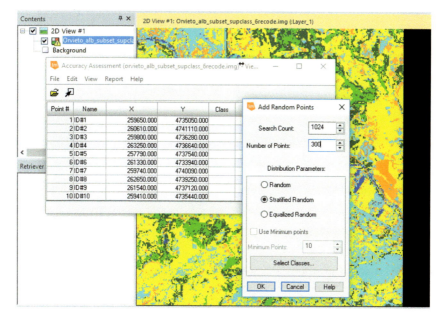

FIGURE 13.3
Accuracy assessment procedure setup for creating stratified random points. (From ERDAS IMAGINE®/Hexagon Geospatial.)

5. Next, click on "Select Classes" in the Add Random Points dialog window and select all the classes (such as 1–6) except "Class 0" (use Shift-Click to select multiple classes). This will exclude the Unclassified Class row in the raster attribute table for the accuracy assessment calculations (Figure 13.4). Finally, close the Select Classes dialog window and select "**OK**" in the Add Random Points dialog window to dismiss these windows.

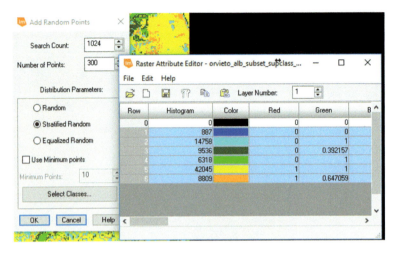

FIGURE 13.4
Selection of classes 1–6 to exclude stratified random points being added to class 0 (unclassified class). (From ERDAS IMAGINE®/Hexagon Geospatial.)

6. Next select **View | Show All** in the Accuracy Assessment tool window. At this point, the 300 hundred reference points are displayed throughout the entire image (Figure 13.5).

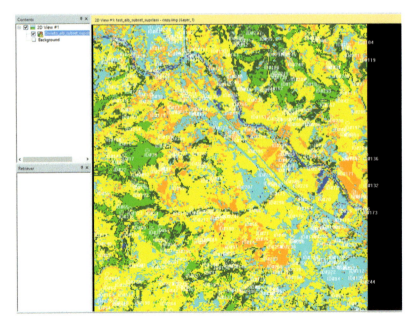

FIGURE 13.5
Output result of stratified random points added to the recoded thematic layer.

Assessing Thematic Classification Accuracy 281

NOTE: You may wish to change the colors of the reference points being displayed in the Accuracy Assessment tool window displayed to make them easier to see if you wish (**View | Change Colors**).

7. In the Contents Legend list, uncheck the recoded supervised classification image (bottom image) so that only the unclassified, Sentinel-2 subset image will be visible in the 2D View window with the 300 hundred reference points displayed (Figure 13.6).

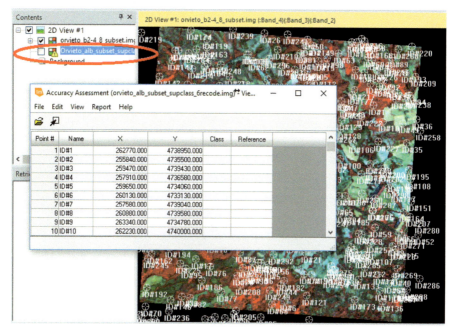

FIGURE 13.6
Initiation of accuracy assessment on the displayed reference image. (From ERDAS IMAGINE®/Hexagon Geospatial.)

8. Next, you will need to input reference points, one by one, into the Accuracy Assessment window's table by doing the following:
   1. In the Point # column, select the row of the first point to enter.
   2. In the **View** menu of the Accuracy Assessment window, select **Show Current Selection**. This will highlight the corresponding Point # on the image in the 2D View window.
   3. Once you have determined the correct land cover category for the corresponding point, enter the number (e.g., 1—water, 2—developed, 3—evergreen, 4—deciduous, 5—agriculture/herbaceous, or 6—barren) into "**Reference**" column.

4. Continue until all rows within the Accuracy Assessment window are complete. Feel free to use the Zoom In and Zoom Out tools within the image to identify the reference class numbers in the Accuracy Assessment window. Continue until all rows within the Accuracy Assessment window are complete (Figure 13.7).

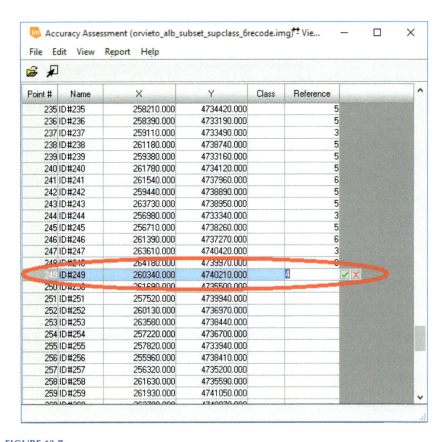

FIGURE 13.7
Input of reference points into the accuracy assessment table window. (From ERDAS IMAGINE®/Hexagon Geospatial.)

Assessing Thematic Classification Accuracy

NOTE: Once you identify the reference land cover class on the image and enter the appropriate number in the Reference column within the Accuracy Assessment window, you will notice that the corresponding reference point in the 2D View window will turn yellow to show that it has now been completed. If you have changed the default reference point colors (**View | Change colors**) in the Accuracy Assessment window's menu, the point colors will follow the options you selected. Also, while determining the reference land cover class on the reference image, you should not look back at the thematic classified image. This will bias the classification results. Only use the **reference image** and your **finely-tuned image interpretation skills** for identifying the following classes represented in Table 13.1.

TABLE 13.1

Land Cover Classification Class and Class Descriptions for the Use in the Land Cover Classification of the Town Orvieto, in the Province of Terni, Italy

Land Cover Classification	Classification Description
1. Water	Contained lakes and small ponds and all other natural and artificial surface waters outside of the USGS National Hydrography Dataset shapefile.
2. Urban/ Development	Contained industrial, commercial, and private building, and the associated parking lots, a mixture of residential buildings, streets, lawns, trees, isolated residential structures or buildings surrounded by larger vegetative land covers, major road and rail networks outside of the predominant residential areas, large homogeneous impervious surfaces, including parking structures, large office buildings, and residential housing developments containing clusters of cul-de-sacs.
3. Evergreen Vegetation	Contained large homogeneous vegetative land covers of trees or shrubs that keep their leaves throughout the winter (mostly coniferous species).
4. Deciduous Vegetation	Contained large homogeneous vegetative land covers of trees or shrubs that lost their leaves during the winter (mostly hardwood species).
5. Agriculture/ Herbaceous Vegetation	Contained agricultural field, grass fields, and urban grasses. Characteristics of this class include large agricultural field, mowed/maintained lawns, fields, and vegetated road medians.
6. Barren Land	Contained areas or fields with little or no vegetation. Characteristics of this class include fallow agricultural fields, bare sediment or soil areas, and areas cleared of vegetation for construction.

Source: Hester D.B. et al., *Photogramm. Eng. Remote Sensing*, 74, 463–471, 2008.

284   *Image Processing and Data Analysis with ERDAS IMAGINE®*

NOTE: You may also find it easier to **zoom** into one point at a time (**View | Show Current Selection**). Also, if you choose, you may want to display the reference image in **True Color** for ease of interpretation.

9. After you have completed entering "**ALL**" of the reference points in the "**Reference**" column, may use want to get an idea of how well the reference class choices compare to the classified image by using the: **Edit | Show Class Values** option.

NOTE: This should only be used after completing all the reference points, as a method of deciding whether you are ready to generate an accuracy assessment report, or if you should try re-doing the exercise altogether. You should not use this option to edit the reference points.

10. Generate report: **Report | Accuracy Report...** The following information will be generated in the report: **Error matrix** to show class accuracies, **Producer's Accuracy** estimate per class, **User's Accuracy** estimate per class, and **Overall Classification Accuracy Overall Kappa Statistic**.

IV. Accuracy Assessment Report Generated from ERDAS IMAGINE

The accuracy assessment report generated (within the Accuracy Assessment window: **Report | Accuracy Report...**) will consist of three main components; an error matrix, accuracy totals, and the Kappa statistics. Each section provides important statistics on the representation of the thematic classification in relation to the reference image.

The error matrix consists of a table that compares the reference image counts (across the top) that represents the *"true"* points in relation to the category on the thematic image. Thus, the columns represent the point counts that are assumed correct and the total number of points correctly identified as true in the reference image. The rows represent the classified thematic image and the total number of points correctly classified in comparison to the reference image. Finally, the counts in the diagonal represent the agreement between classified thematic image and the reference image.

The next section of the accuracy assessment report is the accuracy totals. The accuracy totals the number of counts correctly identified between the classified thematic image and the reference image, and between a Producer's Accuracy and User's Accuracy (see description earlier in the overview section of this chapter). Additionally, this section of the accuracy assessment report provides an estimate of the overall classification accuracy. The overall classification accuracy is expressed as a percentage and is useful for comparing

Assessing Thematic Classification Accuracy 285

the final completed classification to other completed classifications that use similar protocols for achieving a final classified thematic image.

The final section of the accuracy assessment report is the Kappa statistics. The Kappa statistic, or Choen's Kappa, is an estimation of the percentage of classification agreement. This statistic estimates the proportionate reduction in error of a classification compared to achieving at the same result totally by chance (Congalton 1991). process is avoiding 77% of the error that a randomly generated classification would provide (ERDAS 2010, Congalton 1991). The Kappa statistic is given by the following equation:

$$\hat{K} = \frac{P_o - P_{expected}}{1 - P_{expected}}$$

where:

P_o is the Proportion Observed correct: diagonal/classified image total

$P_{expected}$ is the Probability Expected correct: sum of individual class reference totals, multiplied by the individual classified image totals/ Classified Image total

The generated accuracy assessment report for the thematic classified image will resemble the following output (three sections; Error Matrix, Accuracy Totals, and Kappa Statistics):

```
ACCURACY ASSESSMENT REPORT SECTION 1

                        ERROR MATRIX

                       Reference Data
```

Classified Data	Background	Class 1	Class 2	Class 3
Background	0	0	0	0
Class 1	0	3	0	0
Class 2	0	3	48	3
Class 3	0	0	1	31
Class 4	0	0	0	2
Class 5	0	0	1	12
Class 6	0	0	1	0
Column Total	0	6	51	48

Reference Data

Classified Data	Class 4	Class 5	Class 6	Row Total
Background	0	0	0	0
Class 1	0	0	0	3
Class 2	0	0	0	54
Class 3	2	1	0	35
Class 4	21	0	0	23
Class 5	17	120	3	153
Class 6	2	1	28	32
Column Total	42	122	31	300

End of Error Matrix

ACCURACY ASSESSMENT REPORT SECTION 2

ACCURACY TOTALS

Class Name	Reference Totals	Classified Number Totals	Number Correct	Producers Accuracy	Users Accuracy	
Class 0	0	0	0	---	---	(Background)
Class 1	6	3	3	50.00%	100.00%	(Water)
Class 2	51	54	48	94.12%	88.89%	(Developed)
Class 3	48	35	31	64.58%	88.57%	(Evergreen)
Class 4	42	23	21	50.00%	91.30%	(Deciduous)
Class 5	122	153	120	98.36%	78.43%	(Herbaceous)
Class 6	31	32	28	90.32%	87.50%	(Barren)
Totals	300	300	251			

Overall Classification Accuracy = 83.67%

ACCURACY ASSESSMENT REPORT SECTION 3

KAPPA $(\hat{K})$ STATISTICS

Overall Kappa Statistics = 0.7736
Conditional Kappa for each Category.

Class Name	Kappa	
Class 0	0.0000	(Background)
Class 1	1.0000	(Water)
Class 2	0.8661	(Developed)
Class 3	0.8639	(Evergreen Vegetation)
Class 4	0.8989	(Deciduous Vegetation)
Class 5	0.6365	(Agriculture/Herbaceous)
Class 6	0.8606	(Barren)

Review Questions

1. An important component of any land use/land cover classification is the assessing of the accuracy of the classified data. Why is this so important?

2. What is the most widely accepted method for the accuracy assessment of remote-sensing-derived maps?

3. In general terms, discuss the generation of an accuracy assessment.

4. The error matrix produces two types of accuracy estimates. The first of these is referred to as *Producer's Accuracy*. Provide an explanation of Producer's Accuracy.

5. The second accuracy type is known as the *User's Accuracy*. Provide an explanation of User's Accuracy.

14

Basics of Digital Stereoscopy

Overview

An additional procedure for extracting information from imagery is being able to view the relief of the terrain in a simulated three-dimensional view, known as anaglyph images. These types of images are particularly useful in mountainous areas, or regions with variable elevation grades. For example, anaglyph images can be used to study or display the terrain of an area with high a steep or exaggerated relief. Also, anaglyph images can be useful in aiding in the delineation of features within an area of exaggerated relief (such as a mountainous areas). Stereo Analyst is a digital stereoscopic software module within the ERDAS IMAGINE® package. The module allows you to create anaglyphic images from your data that appear three-dimensional when viewed with the red and blue glasses (anaglyphic glasses), which you may have worn in the older 3-D movie houses. Although anaglyphic images are not really a spatial dataset, these types of images can be very informative—as long as the stereo effect is sharp. Imagery and data for this exercise were adapted from original research developed by Dr. Heather Cheshire at North Carolina State University.

Basics of Digital Stereoscopy Application

Learning Objectives

1. To familiarize you with some of the tools, buttons, and menus of working in Stereo Analyst.
2. To determine how to *find* stereo on a computer screen.
3. To create anaglyphic images as an output of your work.
4. To expose you to the delineation of features, as well as how to operate the *floating cursor* of Stereo Analyst.

289

Data required: For this exercise, you will need the following files:

1. **11199_91.img** (and optionally the associated pyramid file: 11199_91.rrd)
2. **11199_92.img** (and optionally the associated pyramid file: 11199_92.rrd)
3. **dupont_91_92.blk**

Additionally, you will also need to obtain a pair of anaglyphic glasses for this exercise. These glasses are readily available, and relatively inexpensive, through vendors such as Amazon or other online retailers (Figure 14.1).

FIGURE 14.1
Basic anaglyphic glasses.

Configuring the Stereo Analyst module

Launch **Stereo Analyst** by opening ERDAS IMAGINE. When the main Imagine toolbar pops up, click on **Toolbox | Stereo Analyst**. This should launch Stereo Analyst in a new window (Figure 14.2).

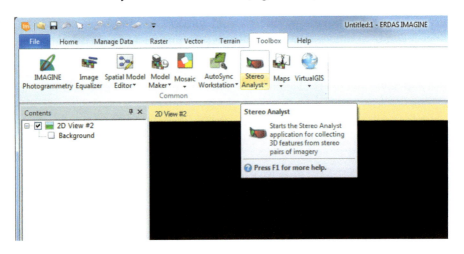

FIGURE 14.2
Stereo Analyst toolbox option. (From ERDAS IMAGINE®/Hexagon Geospatial.)

Basics of Digital Stereoscopy

Before proceeding any further, you must make sure Stereo Analyst is configured appropriately for this exercise. From the **Utility** drop-down menu in Stereo Analyst viewer, select the bottom choice, **Stereo Analyst Options**. A dialog box should open (Figure 14.3).

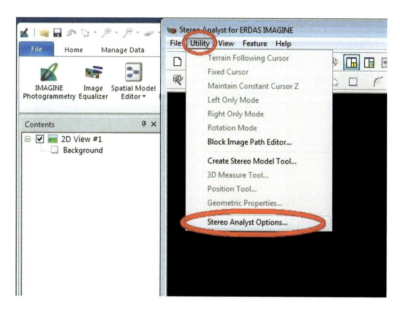

FIGURE 14.3
Stereo Analyst utility options. (From ERDAS IMAGINE®/Hexagon Geospatial.)

Choose **Stereo Mode** from the list of Option Categories. Make sure the Stereo Mode is set to **Color Anaglyph Stereo**, the Left Anaglyph Color Mask is set to **Red**, and the Right Anaglyph Color Mask it set to **Green + Blue**. Once this is **OK**, you can close the dialog (Figure 14.4).

292   *Image Processing and Data Analysis with ERDAS IMAGINE®*

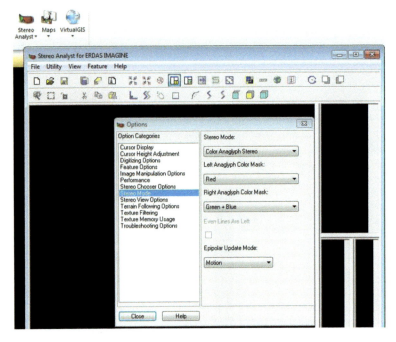

FIGURE 14.4
Stereo Analyst Stereo Mode setup. (From ERDAS IMAGINE®/Hexagon Geospatial.)

NOTE: It is strongly encouraged that you look through the Stereo Analyst online help files, which can be accessed from the **Help** drop-down menu. This help documentation will tell you a great deal about the various keys and menus that will be used for this software.

For this exercise, you should refer to **Navigation Help**, a sub-category that is available separately from the **Help** drop-down menu. On the **Navigation Help** page, you will find basic descriptions of how to navigate with the mouse in Stereo Analyst, and how some of the buttons on the main (or Stereoscope) toolbar work. Before proceeding, review this page. While some of the explanations are a bit obscure, they should assist you in getting around for this exercise. In particular, you will need to know about **Roaming**, **Zooming**, **Rotating**, **Adjusting Cursor Height**, as well as **Toggle Roaming/Zooming/Rotating Left Image Only** and **Toggle Roaming/Zooming/Rotating Right Image Only**.

Basics of Digital Stereoscopy

Making a Digital Stereo Pair

You will start by creating a stereo pair dataset with two overlapping images from Dupont State Recreational Forest in western North Carolina. For more information about the forest, check out the following website: http://ncforestservice.gov/Contacts/dsf.htm

1. In Stereo Analyst, choose **Open** from the File drop-down menu, then select **Open an Image in Mono**. Navigate to the appropriate folder. Select **11199_91.img** and click **OK**.

2. Again, choose **Open** from the File drop-down menu, this time select **Add a Second Image for Stereo**. Navigate to the appropriate folder. Select **11199_92.img** and click **OK**.

   Now you will have two images in the main window, one that is red (left) and one that is blue (right). The images also appear in a window on the top-right of the Stereo Analyst interface, however we will concentrate on the images in the main window for now.

3. Now put on the anaglyphic glasses. The red lens goes over the left eye.

4. If you look closely at the two images, you may notice that there are areas in each image that will overlap the other image. This makes a stereo pair or a Digital Stereo Model (DSM) possible (Figure 14.5).

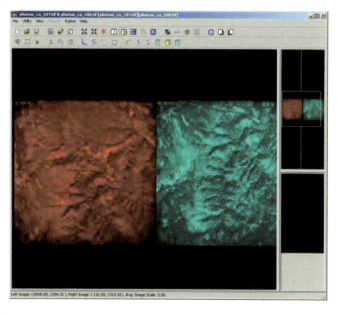

FIGURE 14.5
Stereo Analyst Stereo Mode with two images in the main window, one that is red (left) and one that is blue (right). (From ERDAS IMAGINE®/Hexagon Geospatial.)

5. You can manipulate a single image by choosing the **left-hand image** button to move just the left image, or the **right-hand image** button to move just the right image. When you click one of those two buttons to activate it on the toolbar, then click on the appropriate image using the mouse (and while holding down the left mouse button), you can drag the image into place so that they overlap (Figure 14.6).

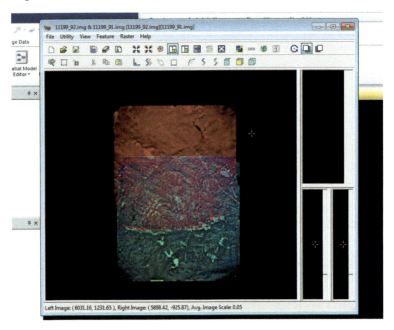

FIGURE 14.6
Manipulate a single image options (left-hand image option selected). (From ERDAS IMAGINE®/Hexagon Geospatial.)

6. Next rotate both images so that the overlap goes left-to-right, not up-to-down.

7. While it is possible to rotate each of the two images independently, it is easier to rotate them together. Click on the **Rotation** button to activate it, and then double-click within the mouse in the main window—just above the two images. A crosshair should appear.

8. Make sure the cursor is inside the crosshair, and then hold down the left mouse button. Now, by moving the mouse you can rotate the entire view (including both images). If only one image is rotating, check to make sure neither of the image selection buttons are activated.

Basics of Digital Stereoscopy 295

9. Rotate the images 90° counter-clockwise. The annotation on the top of the two images should be on the left-hand side (Figure 14.7).

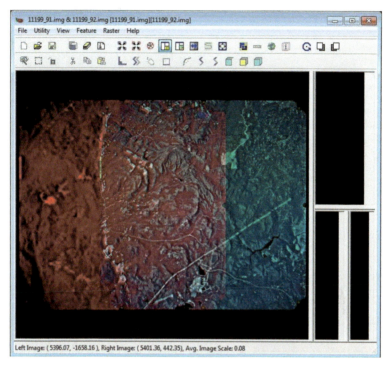

FIGURE 14.7
Images rotated 90° (counter-clockwise). (From ERDAS IMAGINE®/Hexagon Geospatial.)

10. Click on the **Rotation** button again to deactivate it.
11. Now move the images into place to create a stereo effect. Using the left and right image toggle buttons (again, click on a button to activate it), move one or both images so the images overlap (Figure 14.8).

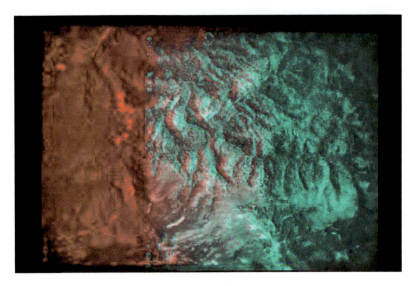

FIGURE 14.8
Positioning of the red image on the left side and the blue image on the right side to create a stereo effect. (From ERDAS IMAGINE®/Hexagon Geospatial.)

> **NOTE:** Be sure to maintain the **red image on the left side** and the **blue image on the right side** (Figure 14.9).

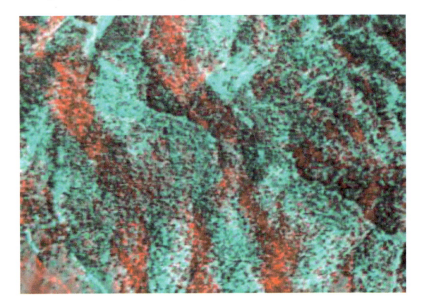

FIGURE 14.9
Sample stereo image effect visualized while wearing anaglyphic glasses. The landscape topography/relief should be visible throughout the image. (From ERDAS IMAGINE®/Hexagon Geospatial.)

Basics of Digital Stereoscopy

12. Once you have the images in the correct positions, you should start to see a stereo effect. Remember, only the area where the images overlap will appear in stereo.

    NOTE: Be sure to deactivate the toggle buttons after you are done. It is very tempting to immediately try to zoom closer once you have basic stereo established. You don't want to accidentally zoom in on one image and not the other—that will certainly get rid of the stereo effect!!!

    This is a sample stereo image, which you should be able to see while wearing anaglyphic glasses.

13. Once you have a stereo display, zoom in fairly tightly on the image. (To zoom, hold down the mouse wheel—don't roll it—and then push the entire mouse forward, away from you. To zoom out, hold down the wheel and pull the entire mouse backward, towards you. There are other zooming methods available in the help files, such as holding down **CTRL + Lt Mouse** and pushing the mouse forward or backward, and others)

14. Can you still see a strong stereo effect? If not, you may again move the images independently as previously described—sometimes a slight tweaking is necessary to make sure the images are lined up appropriately.

15. Once the stereo effect seems good, you may proceed to the next part of the exercise. If you would like to save the stereo setup, you may do that as well. From the **File** drop-down menu, select **Export**, and then select **Relative Stereo Pair**. A dialog will open (Figure 14.10).

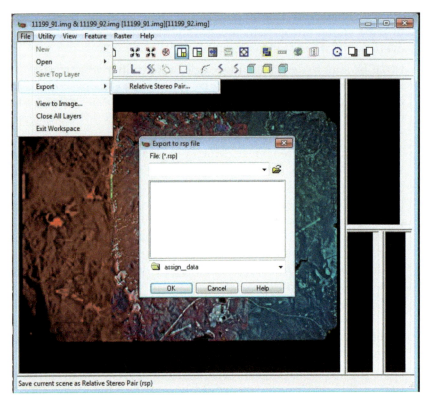

FIGURE 14.10
Save the Digital Stereo Model setup as a Relative Stereo Pair (*.rsp file). (From ERDAS IMAGINE®/Hexagon Geospatial.)

16. Save the file **dupont.rsp** to the working folder where all the images are stored. This is a pointer file that stores the orientation of the two images as you have just arranged them. Now, you could close Stereo Analyst and continue from this point later, by choosing **Open Relative Stereo Pair** from the **File** drop-down menu.

Creating Anaglyph Images for Export

Unfortunately, there is no particularly straightforward way to export an anaglyph image from within Stereo Analyst. Using the **View to Image** option under the **File** drop-down menu, you do have the option to export whatever is displayed in the main Stereo Analyst view window to an Imagine or TIFF format file. While these can be configured and/or exported to JPEG format, and

Basics of Digital Stereoscopy

so on., through the main Imagine interface, it is not a particularly easy process. The following describes a simple alternate method for extracting anaglyphs.

1. While wearing the anaglyphic glasses, move the stereo pair or zoom as necessary so that the area of interest fills the main view window in Stereo Analyst. You will make one anaglyph from the photography at an **Image Scale of 0.20**. The image scale is displayed on the bottom left-hand bar of the Stereo Analyst interface (Figure 14.11).

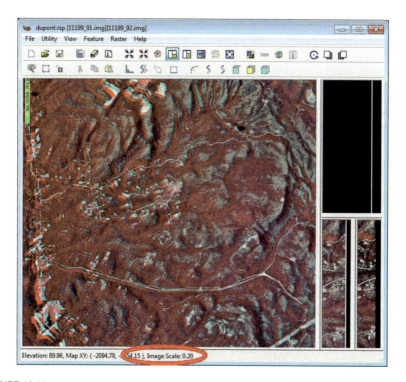

FIGURE 14.11
The image scale (0.20) is displayed on the bottom left-hand bar of the Stereo Analyst interface. (From ERDAS IMAGINE®/Hexagon Geospatial.)

2. Make sure Stereo Analyst is the active application on your computer. Press **Ctrl + Alt + Print Screen** (simultaneously) on your keyboard.
3. Return to your computer's main **Start/Programs** menu, locate the **Accessories** folder, and select the Microsoft Paint program ("**Paint**"), or simply search for the Paint program in the Windows search bar.

300   *Image Processing and Data Analysis with ERDAS IMAGINE®*

4. When Paint opens, select **Paste** from the **Edit** menu. You may get a message box indicating the image in the clipboard is larger than the bitmap. Select **Yes** to enlarge it.

5. Activate the **Crop** tool but clicking on the appropriate button on the toolbar.

6. Draw a box around the area that you want to select from the output image. This probably means clipping everything (the toolbar, side windows, etc.) except the central part of the Stereo Analyst image.

7. Once you have drawn a box around the correct image area, click on **Crop** again. This should clear everything from the Paint window except the area of interest.

8. From the **File** drop-down menu, choose **Save As**. The 24-bit bitmap image will probably offer the best image quality via the Paint program, but it also yields a large file. A JPEG file tends to be a sufficient compromise between file size and image quality.

   NOTE: You may also choose to use any other screen capture software, such as Adobe Photoshop.

Delineating in Stereo

You will now attempt to delineate some features from images of the Dupont State Park. These images were created from hard-copy photographs, at a scale of 1:40000. The following instructions present some of the more basic functions within Stereo Analyst, however additional information can be obtained by consulting the Stereo Analyst help files.

1. To start with, close all currently open layers in Stereo Analyst (this option is in the **File** drop-down menu). Then select **Open** from the **File** menu, choosing to open a **Block File**. Navigate to the folder with all the newly created files. There should be a file named **dupont_91_92_blk**. (the file type may need to be changed to Block File). Open this file.

2. You may receive the following warning (Figure 14.12):

Basics of Digital Stereoscopy 301

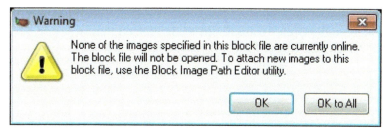

FIGURE 14.12
Warning that Block Image Path Editor must be used to attach the individual images to the block file. (From ERDAS IMAGINE®/Hexagon Geospatial.)

3. Click **OK**. Then Click on **Utility** in the file menu and then **Block File Image Path Editor**. Left-Click on the **RED** block in the Online column and navigate to the appropriate file, and then click **OK**. Do this for all missing files until each square in the Online column turns **GREEN**, and then click on Save in the **Block File Image Path Editor** dialog (Figure 14.13).

FIGURE 14.13
Block Image Path Editor used to attach the individual images to the block file. (From ERDAS IMAGINE®/Hexagon Geospatial.)

4. Continue editing the image paths for all missing files until each square in the Online column turns **GREEN**, and then click on **Save** in the **Block File Image Path Editor** dialog (Figure 14.14).

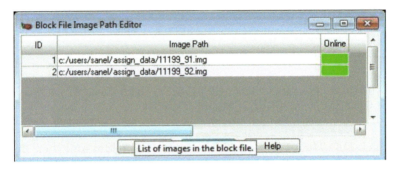

FIGURE 14.14
Block Image Path Editor showing that the individual images are now attached to the block file. This attachment is signified by green color in the Online column. (From ERDAS IMAGINE®/ Hexagon Geospatial.)

5. Now open, open the file named **dupont_91_92_blk** as described in step #1. You will have a pair of images in stereo. (Obviously, you will need the anaglyphic glasses to see this.)
6. Go to **File | New | Stereo Analyst Feature Project**. Under the **Overview** tab, navigate to your file directory and name the new project **Drainage** (file will be **Drainage.fpj**).

NOTE: Do NOT click OK until completing all **Feature Project** setup steps.

7. Click on the **Feature Classes** tab (**feature | feature project properties | feature classes**). You will be delineating a drainage basin and drainages on the stereo image, so you will need a polygon class and a line class. Create two feature classes using the available categories. (Suggestion: **Under Rivers, Lakes and Canals**, you might use **Dry Lake** for the basin and **Perennial Stream** for the drainages.) (Figure 14.15).

Basics of Digital Stereoscopy

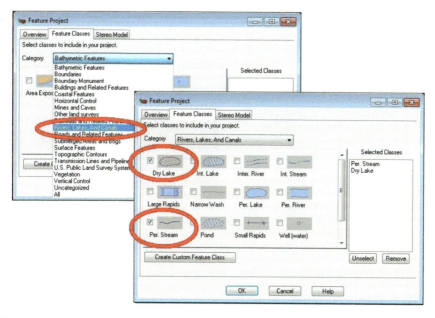

FIGURE 14.15
Selection of feature class attributes as features to delineate. (From ERDAS IMAGINE®/Hexagon Geospatial.)

8. In the **Feature Project** dialog box, make sure the streams and basins classes you just created show up in the **Selected Classes** sections as illustrated earlier.
9. Under the **Stereo Model** tab, make sure **dupont_91_92.blk** is selected under "**Current Image for Feature Collection**," then click **OK** (Figure 14.16).
10. Tools for creating a basin polygon and a stream should appear in the left margin of the stereo viewer (Figure 14.17).

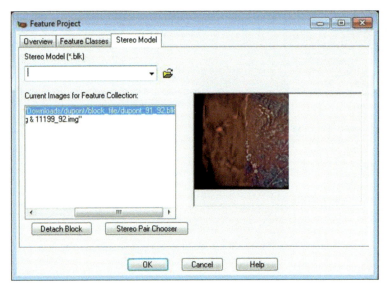

FIGURE 14.16
Selection of dupont_91_92.blk as "Current Image for Feature Collection." This indicates that these block file images will be used for delineation. (From ERDAS IMAGINE®/Hexagon Geospatial.)

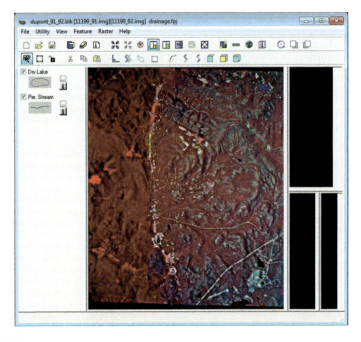

FIGURE 14.17
Polygon delineation tool (i.e., Dry Lake) and line delineation tool (i.e., Perennial Stream) displayed to the left of the stereo viewer. (From ERDAS IMAGINE®/Hexagon Geospatial.)

Basics of Digital Stereoscopy   305

11. You can change the display properties of the class by clicking on the button above the *i* (...) button.
12. Use these tools to delineate the drainage basin and stream with the area indicated in the following photo (Figure 14.18).

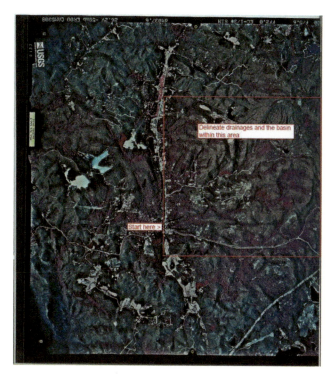

FIGURE 14.18
Lake and stream areas to be used for delineations.

13. Once you select each feature class you should be able to select from several mouse functions to help you delineate. These include:

    Click on a **Feature Button** (such as the button for "dry lake"). Move the cursor inside the stereo image block. Make sure the cursor is resting on the ground (use the scroll wheel on the mouse—watch the cursor move up and down).

    - A single **left-click** will mark a vertex.
    - Continue delineating around the basin.
    - **Double-click** to close and end the polygon.

14. Repeat a similar procedure for **Perennial Stream** (Figure 14.19).

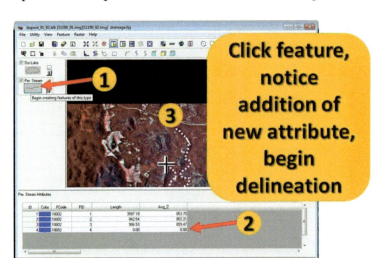

FIGURE 14.19
Steps for line delineation of a perennial stream. (From ERDAS IMAGINE®/Hexagon Geospatial.)

Review Questions

1. What are anaglyph images?
2. In what type of terrain areas, or regions of the landscape might anaglyph images be particularly useful?
3. In what other ways might anaglyph images be useful, with respect to feature extraction through the process of delineation?
4. Delineated features from an anaglyph images are stored as what type of data? Why is this an advantage?
5. How are anaglyph images created and used in image processing packages?

Appendix: Answer to Chapter Review Questions

Answers to Chapter 1 Review Questions

1. Remote sensing can be characterized as being the science of acquiring any information about an object or a phenomenon on the surface of the Earth without coming into physical contact with it. Typically, with remote sensing image analysis from aircraft or satellite sensors, light energy as it relates to the electromagnetic spectrum, is captured and used to form *images*.

2. Data captured from remote sensing of earth features are also typically based on four types of resolutions. These resolutions are known as spatial, spectral, radiometric and temporal resolutions.

3. Electromagnetic radiation (EMR) is defined as all energy that moves with the velocity of light in a harmonic wave pattern. The harmonic wave pattern refers to the waves that are equally and repetitively spaced in time. Some components of the electromagnetic spectrum consist of visible light, the portion of the EMR that humans are able to see, as well as radio waves, microwaves, RADAR, infrared, ultraviolet, X-rays, and gamma rays.

4. The Landsat 8 Operational Land Imager (OLI) and Thermal Infrared Sensor (TIRS) or simply Landsat 8, was launched on February 11, 2013, and represents the United States' continuing mission of the Landsat sensor program. The sensor's images consist of nine spectral bands (Bands 1–7, and 9) with a spatial resolution of 30 meters (m), one panchromatic band (Band 8) with a spatial resolution of 15 meters, and two thermal, surface temperature, bands (Bands 10 and 11) with a spatial resolution of 100 meters (resampled to 30 meters). Landsat 8 has a revisit cycle of every 16 days.

5. The Sentinel-2 multispectral sensor is a component of the ESA's Copernicus Programme. The initial Sentinel-2A satellite was launched June 23, 2015. This was followed up by a Sentinel-2B satellite launch on March 7, 2017. The two Sentinel satellites (2A and 2B) operate in a 180° phase with the orbit allowing for global coverage of the Earth's surface every five days, or less at higher latitudes (every 10 days with one satellite). The Sentinel sensors' images consist of

308   *Appendix*

13 spectral bands and operate in the visible/near infrared (VNIR) and shortwave infrared spectral range (SWIR). The spectral bands possess a mix of spatial resolutions (10 m, 20 m, and 60 m).

Answers to Chapter 2 Review Questions

1. True: A number of digital image analysis packages are currently available. Some will be more feature packed and user-adaptable than others. However, any investigator should analyze their current available options as it relates to their own specific needs.

2. True: Freeware and open source packages are becoming more available and highly competitive to many of the commercial packages.

3. True: USGS EarthExplorer (http://earthexplorer.usgs.gov/), operated by the US government maintains a highly comprehensive archive of raster-based, remotely sensed data from all over the world.

4. The Show Metadata and Browse icon is used to determine whether the recorded image quality meets the specified criteria necessary. For example, an estimate of the amount of cloud cover in the scene is given. Having an idea of the amount of cloud cover present within the scene is important in estimating the possibility of ground features being obscured by clouds within the area of interest.

5. The types of information that may typically be available from reviewing the image metadata in most digital image processing software packages would include the number of bands that the image has, as well as the maximum, minimum, and mean pixel values for each image band. Additionally, information such as spatial reference (i.e., projection, spheroid, and datum), file location of the image source, and the number of rows and columns of pixels in the image, as well as statistical or histogram plots may also be available.

6. The appearance of an image depends on how you choose to assign the three additive primary colors (red, green, and blue) to satellite image bands. Each image spectral band has captured a different part of the electromagnetic spectrum. By assigning colors that represent a "True Color" or natural color image, means that the image shows objects in the same colors that your eyes would normally see (e.g., grass is green, and water is blue). A "False Color" image means that the colors have been assigned to three different wavelengths that your eyes might not normally see (i.e., vegetation appears in shades of red).

Appendix 309

7. When creating a multiband using a layer stack operation, only the single bands that possess similar spatial resolutions should be used. In the layer stack operation all bands in the resulting multi-band image would be resampled to the smallest pixel-meter size. This would result in the loss of some spectral definition in the bands with larger, original, spatial resolutions and potentially provide unreliable results if these bands are used in classification.

8. The "**Swipe**" button ▨ Swipe , introduced in the "Opening Multiple Images in a Single Window" section of Chapter 2, provides some very useful functionality. For example, this tool can be useful when comparing imagery of the same location at different times (seasons, years, etc.).

Answers to Chapter 3 Review Questions

1. Geometric image distortions are generally divided into two types of errors, systematic and nonsystematic errors. Systematic distortions are due to image motion caused by the forward movement of the spacecraft, variations in the mirror scanning rate, panoramic distortions, variations in platform velocity, and distortions due to the rotation as well as the curvature of the Earth. These types of errors may be compensated for, reduced, or corrected, on board the aircraft or satellite, or through preprocessing using standard equations. Because these types of errors are assumed to be systematic (roughly affecting the entire image equally), an applied correction procedure will systematically function to adjust all pixel locations throughout the image. Nonsystematic distortions are typically due to variations in satellite altitude, speed and angular orientation in reference to the ground (attitude) and may be more difficult to identify and remove or reduce.

2. The georectification processing application introduced two georectification models, the polynomial model and the rubber sheeting model. Each operation provides some advantages and disadvantages when considering the final georectified image. For the polynomial model, possible advantages and disadvantages that may be noticeable in the final georectified image, include that this model allows for the full area of the input image to be georectified and displayed. However, the disadvantage here is that some areas within the image may likely have error in the alignment.

3. For the rubber sheeting model, possible advantages and disadvantages, which may be noticeable in the final georectified image,

310   *Appendix*

include that this model allows for a very accurate georectification but may clip or *trim* away the edges of the input image in an effort to get the best possible fit. The disadvantage here is that the trimmed away edge data represents a loss of data that may be been useful for the intended study.

4. A situation where the polynomial model might be beneficial over the rubber sheeting model may be when georectifing an image or photo with lots of anthropogenic features that were spatially distributed throughout the entire raw image and would expect to be distributed throughout the final georectified image. This is typically the case with in urban imagery.

5. A situation where the rubber sheeting model might be beneficial over the polynomial model may be when georectifing an image or photo that very high accuracy of the final georectified image is required. Additionally, it would be helpful that the input image is much larger than the area of interest, so any clipped or trimmed edge portions of the input image would not impact the integrity of the intended study area.

Answers to Chapter 4 Review Questions

1. Orthorectification differs from polynomial transformations and rubber sheeting in georeferencing an uncorrected digital image to a reference image in that it uses z-coordinate values (elevation data) in the georeferencing process. The use of elevation data in orthorectification allows for the correction of relief displacement. Orthorectification also incorporates camera-specific information to correct certain optical displacement effects.

2. The output of orthorectification is an orthophoto that is described as planimetrically accurate. The term planimetrically accurate in an orthophoto refers to the scale being constant across the image.

3. An advantage to having a planimetrically accurate image or orthophoto would allow the orthophoto to be used as a base map, which are commonly used in a wide variety of geographic information systems (GIS) and geospatial applications.

4. In creating an orthophoto in a digital image processing package, it is important that each ground observations, or ground control point (GCP), must be selected from features that can be clearly seen in both images. This is to ensure that the input image aligns correctly with the reference image.

Appendix 311

5. Typical features used for GCPs include edges or intersections of roads, parking lots, and so on—these features tend to be the most easily discernible on black-and-white DOQQs or similar digital images. Man-made features are often used, but occasionally single trees in the middle of agricultural fields or lawns can serve as an acceptable ground control.

Answers to Chapter 5 Review Questions

1. Positional accuracy refers to the spatial agreement of an object displayed on an image, compared to the actual position of that object on the ground (true position). Classification accuracy refers to the characterization of, typically, raw pixel data into categories representing defined land use and land cover groupings or classes, and how well these classes agree with reference data.

2. Every image is potentially distorted in the X or Y direction, or both. In other words, the coordinate space of an image is a projection. If the *projection* of the image is compatible with the projection of the data that is being used as a reference image (i.e., actual ground positions), then the Root Mean Square (RMS) is dependent on how closely the "From points" in the image relate to the "To points" in the data (i.e., actual ground positions).

3. The term Root Mean Square Error (RMSE) is defined as a measure of positional accuracy. The RMSE (or RMS error) is the absolute fit of the corrected positional accuracy transformation model to the data, or the difference between values predicted by the transformation model and the actual values observed.

4. The RMSE value is calculated by computing the difference between each predicted value of the transformation model and the actual observed values, squaring these differences, calculated the sum of the squared differences divided by the total number of values, and, finally, taking the square root.

5. The Total RMSE refers to the total root mean square (RMS) error for an entire image. The Total RMSE takes into account the individual RMS errors for each observation or ground control point (GCP). In general terms, the RMSE is a measure of how much distortion, or stretching, has occurred in the image (in the X or Y directions, or both) that is being georeferenced. The farther out of shape the image becomes, the larger the RMS. Although in some cases, a large RMS will occur when an image is severely distorted or there is a need for greater accuracy along one edge of the image.

312   *Appendix*

Answers to Chapter 6 Review Questions

1. The radiometric resolution refers to the dynamic range, or a number of tonal levels, at which data for a given spectral band are recorded by a particular sensor.

2. Often it is necessary in most digital image processing packages to apply radiometric enhancement techniques to the raw imagery being displayed. Without the application of radiometric enhancement techniques, the raw data captured by the sensor is too dark to be useful by the image analyst in interpreting features on the ground.

3. Radiometric enhancements work to increase the interpretability of the raw imagery being displayed by performing operations that adjust the contrast, or in some cases stretch the tonal levels of the radiometric range of the data.

4. An image with a 1-bit image radiometric resolution would contain two brightness values, or grayscale levels (black and white), within the data. An image with an 8-bit radiometric resolution would have up to 256 grayscale levels within the data. An image with an 11-bit image may contain up to 2,048 grayscale levels. Landsat 4–7 possess an 8-bit radiometric resolution. The Landsat 8 sensor possesses a radiometric resolution captured in a 12-bit dynamic range (4,096 grayscale levels) and delivered as 16-bit images when processed into Level-1 data products (scaled to 55,000 grayscale levels). The 12-bit dynamic range radiometric resolution of the Landsat 8 sensors is also comparable to Sentinel-2's radiometric range.

5. Assuming an unsigned 8-bit radiometric resolution image (256 grayscale levels) composed of three land cover classes: (1) clear, deep, water/lake; (2) homogenous deciduous vegetation; and (3) homogenous, imperious surface. The water/lake land cover class would be expected to produce dark pixel values.

6. In the theoretical image described in the above question (Question 5), the homogenous, imperious surface land cover class would be expected to produce bright pixel values.

Answers to Chapter 7 Review Questions

1. Spatial enhancements, like radiometric enhancements, also improve the interpretability of features within the data. However, instead of operating on individual pixels as in the radiometric enhancement

Appendix 313

models, the spatial enhancement models operate by modifying neighborhood pixels values based on the value of a targeted pixel.

2. Common spatial enhancement operations include the following: convolution filter, non-directional edge detection, focal analysis, statistical filter, and image degrade.

3. Some common spectral image enhancement techniques include the band combination creation of true color composite images and false color composite images (i.e., vegetation/forested areas displayed as varying shades of red).

4. The convolution filter applies a matrix of small neighborhoods of cells (i.e., 3×3, 5×5, 9×9, etc.) to compute a neighborhood average as the matrix moves throughout the entire image. The result, in the case of a low pass convolution filter, is that the image may appear smoother, or blurred, in appearance (less detailed) based on the filter size.

5. Low pass filters are spatial enhancements that decrease differences between the target pixel and its surrounding pixels. These filters can be useful for removing image noise and variability. High pass filters are spatial enhancements that increase differences between the targeted pixel and its surrounding pixels. A high pass convolution filter may highlight the boundaries, or edges, occurring between homogeneous groups of pixels.

Answers to Chapter 8 Review Questions

1. Digitization is a method of data input that has traditionally required a manual procedure to enter spatial data into a digital database from physical or hard photographs, maps, or other imagery. This method, while very laborious and time-consuming, could also introduce several areas associated with the data import process and the creation of the database topology.

2. Often the database created from manual digitization is stored as vector information. This vector information is often compatible with GIS packages, and can then be used for land use and land cover classification or other land cover analysis operations.

3. Modern GIS and image processing software packages now allow for a *heads-up* digitization method. This method allows the analyst to digitize features from photographs, maps, or other imagery directly from a computer screen using a mouse cursor as the input device. Each mouse click inserts nodes or polygon vertices into the captured vector database.

314   *Appendix*

4. Modern GIS and image processing software also allows for a semi-automated feature extraction technique. This technique allows for faster recognition of boundary features within the image data being digitized. This semi-automated feature extraction technique greatly reduces the amount of labor, times and errors associated with traditional digitizing.

5. The manual digitization process would seem more effective as a data input method for an area with larger, homogenous, and connect features, as the process would require much less mouse clicks when compared to an area of the same size, with smaller, unevenly distributed, and non-contiguous features.

Answers to Chapter 9 Review Questions

1. The term classification, in relation to the conversion of raw imagery into a thematic image, refers to the operation of grouping or sorting image pixels, typically based on the pixel spectral values, into categories (or themes) is known as image classification. In the image classification process, each *classified* category represents a unique land cover class that is representative of identified features on the ground. The resultant image classification map or *categorized data* is known as a thematic map.

2. Two traditional image classification methods are known as unsupervised classification and supervised classification.

3. The unsupervised method of classification typically requires little or no input from the image analyst in developing the output land use/land cover classification. In this method, the classification system, or classifier, uses statistical means and covariance matrices to iteratively assign each pixel to a designated output class based on how spectrally separate each group of clustered pixels are. The analyst is then required to assign labels to the output categorical groups to complete the classification.

4. The Iterative Self-Organizing Data Analysis Technique (ISODATA) utilizes the mean and standard deviation in several bands in n-dimensional space to iteratively group similar pixels. The user will guide the operation by selecting initial parameters that include the maximum number of clusters, maximum percentage of pixels that can remain unchanged between iterations, maximum number of iterations, minimum percentage of pixels assigned to each cluster, maximum standard deviation, and minimum distance between clusters.

Appendix 315

5. The k-means approach represents a second clustering algorithm that also may be used in the generation of an unsupervised classification. This approach also works in an iterative process. The initial stage includes the buildup of a number of similar pixel clusters within a range determined by the analyst. Each resulting cluster is composed of pixel groups consisting of similar spectral values, which likewise occupies a common spectral space that consists of with a well-defined mean vector for each class. This operation is followed by a minimum-distance-to-means classification algorithm that determines the final pixel cluster groups. In the case of the k-means clustering algorithm, the data space is partitioned into Voronoi cells. The Voronoi cells represent regions that each contain sets of points, or seeds, which correspond to the cluster means are closest in multidimensional Euclidean space. In the k-means operation, pixels are iteratively classified into a predetermined number of clusters with no deletion, splitting, or merging of the clusters between each iteration.

Answers to Chapter 10 Review Questions

1. In general terms, the process of a supervised classification involves direct input from the analyst in identifying training samples across the imagery, which represent multiple corresponding areas on the ground and will be used to generate the completed supervised classification.

2. A supervised classification generally consists of three stages performed by the analyst. These stages include training, classification, and the final output.

3. The training stage consists of the initial stage of a supervised classification. In this stage, the analyst inspects the image to be classified and uses knowledge of the area (collected from field visits, reference maps or photos, or other higher quality and higher resolution data) to collect training sites within the imagery that represent the corresponding areas on the ground. The training sites may be collected in the form of delineated polygons or representative pixels that the software will use to develop a multiband classification based on spectral relationships scattered from the sites. Each training site should represent a homogeneous and contiguous grouping of pixels within an individual category of interest. The number of training sites collected from the imagery should also capture the amount of variability contained within the category of interest as identified

across the entire image data. Also, each training site for each category of interest should be randomly or systematically distributed throughout the entire image data.

4. The classification process represents the second stage of supervised classification. This process uses statistical algorithms to analyze the spectral bands of the imagery and determine how closely each relates to the identified training samples representing the categories of interest throughout the entire image data. The statistical algorithms represent the most widely used algorithms in the classification process. These include the minimum distance to means, parallelepiped, and maximum likelihood.

5. The last stage of the supervised classification process is known as the final output stage. The final output may produce a thematic, or classification map, representing each category of interest developed originally from the training samples. The thematic data may further be analyzed to determine land use and land cover areas for each defined category. Additionally, thematic data be imported into other geospatial and statistical tools for further mapping and modeling. Classified data products may also be used for visualizing and presenting land use and land cover analysis results, or even modeled with other temporal land cover data to develop change analysis models. The thematic classification data, along with the accompanying statistical parameters, may also be used to develop an assessment of accuracy to determine an estimate of the thematic classification accuracy.

Answers to Chapter 11 Review Questions

1. Object oriented classification, image extraction, or Object Based Image Analysis (OBIA), seem to be more useful for higher-spatial resolution imagery than the traditional pixel-based approach provided by the supervised or unsupervised classification methods because the traditional approaches have worked well when the identified land cover classes throughout an image exhibit a definable spectral separation or distance. However, more recent higher-spatial resolution imagery (~1–≤1 m), has proved more challenging to use than these traditional approaches due to the increase of spectral variability within the target classes.

2. The OBIA method differs from the traditional pixel-based approach of the supervised or unsupervised classification methods by providing a more automated method of image classification.

Appendix 317

3. In an OBIA approach, in addition to using spectral properties, the software also uses image-based *cues* that are similar to a human's process for visual interpretation. These cues include color/tone, texture, size, shape, shadow, site/situation, pattern, and association.

4. Additionally, the OBIA procedure applies classifications-based objects, which have been established from training samples and that represent the features needed to be classified. These objects are then defined within the software based on rules that are further used to model the individual or groups of objects based on size, shape, direction, distance, distribution throughout the image, texture, spectral pixel values, position, orientation, and relationship between objects, and so on, as other user-defined parameters.

5. After each object has been identified, the OBIA procedure may then complete the classification operation by incorporating statistical algorithms, such as nearest neighbor analysis, neural networks, and decision tree analyses.

Answers to Chapter 12 Review Questions

1. Two additional classification analysis techniques that may be useful in the process of generating an effective land use land cover classification include the creation of water-only binary image mask and a Normalized Difference Vegetation Index (NDVI), also known as a greenness map.

2. The water-only image mask is useful for masking out, or removing, the water from an image. A water-only image mask may be particularly useful in an image where water pixel values encompass a high spectral range.

3. The NDVI it is a popular tool used in forest vegetation health assessments and represents an index of vegetation greenness.

4. A NDVI image is created for specific image dates through the division of the image's red and near infrared spectral bands (i.e., $NIR - RED/NIR + RED$). Chlorophyll pigments (light absorbing pigments) found in the leaves of healthy vegetation absorb incoming radiation (visible light) in the blue (0.45 μm) and red range (0.67 μm) of the electromagnetic spectrum (EMS). Green (0.5 μm), and more so in the near infrared (0.7–1.3 μm), ranges of the EMS light is reflected. Particularly in the near infrared portion of the EMS, there is a strong reflection of light largely due to the internal makeup of the leaf's structure. When vegetation becomes stressed, due to drought, infestation, disease, and so on, the leaves typically decrease reflection in the near infrared range as the leaf's internal structure begins to change.

318   *Appendix*

5. NDVI is useful for identifying areas where vegetation is under stress. Multiple NDVI images can be generated to determine the spectral signatures for areas of healthy vegetation versus areas of stressed vegetation.

Answers to Chapter 13 Review Questions

1. An important component of any land use/land cover classification is the assessing of the accuracy of the classified data. It is important to note that results of a classification or output map represents an imperfect depiction of the original data. All classification outputs contain errors, and it is the responsibility of the remote sensing analyst to characterize these errors prior to a map's use in subsequent applications.

2. The most widely accepted method for the accuracy assessment of remote-sensing-derived maps is by comparison to reference data (also known as *ground truth*) collected by visiting an adequate number of sample sites in the. The key instrument in this comparison is the generation of an accuracy assessment.

3. In general terms, the generation of an accuracy assessment includes an error matrix, which quantifies the accuracy for each map class of interest as well as the overall map accuracy (i.e., combining all of the classes) and the concepts of Producer's Accuracy and User's Accuracy and the Kappa statistics. The accuracy assessment represents the reliability and theoretical repeatability of the final thematic classification generated by your methods. In a land cover classification project, an error matrix is typically produced to demonstrate class accuracies based on your classified categories. Diagonal elements of the error matrices are the number of sites correctly classified in the image. The sum of off-diagonal elements in each column indicates the number of sites not identified as being in a particular class. This type of error is known as an *omission error*. The sum of off-diagonal elements in each row indicates the number of sites identified as being in a particular class when they, in fact, belonged to a different class. This type of error is known as a commission error. Error matrices are useful for determining classification categories, or classes, that are most likely to be confused and are sometimes referred to as *Confusion Tables*.

Appendix 319

4. The error matrix produces two types of accuracy estimates. The first of these is referred to as *Producer's Accuracy*. The Producer's Accuracy is the probability that an area that is in class "X" has been correctly identified as being in class "X." This accuracy is indicative of possible errors of omission as it defines the number of verification sites that were actually *found* in the classification.

5. The second accuracy type is known as the *User's Accuracy*. The User's Accuracy is the probability that an area that has been classified as "X" actually is in class "X." This is indicative of errors of commission as it defines the number of verification sites *committed* to the correct class.

Answers to Chapter 14 Review Questions

1. Anaglyph images are images that provide for the visualization of the relief of the terrain in a simulated three-dimensional view.

2. Anaglyph images are particularly useful in mountainous areas, or regions with variable elevation grades. For example, anaglyph images can be used to study or display the terrain of an area with high a steep or exaggerated relief.

3. Also, anaglyph images can be useful in aiding in the delineation of features within an area of exaggerated relief (such as mountainous areas).

4. Delineated features from anaglyph images are stored as vector data. The resulting vector data has the advantage of also being compatible in most GIS platforms.

5. Anaglyph images are typically created in image processing packages by creating stereo images pairs from generally two overlapping images that are slightly off-set. The software sets up the stereo image pairs by assigning a red filter to one image and a blue filter to the other to create a three-dimensional effect when viewed with anaglyphic glasses.

References

Blaschke, T. 2010. Object based image analysis for remote sensing. *ISPRS Journal of Photogrammetry and Remote Sensing* 65:2–16.

Barr, S., and M. Barnsley. 2000. Reducing structural clutter in land cover classifications of high spatial resolution remotely-sensed images for urban land use mapping. *Computers and Geosciences* 26:433–449.

Blaisdell, E.A. 1993. *Statistics in Practice* (Harcourt, New York), 653 p.

Celik, T. 2009. Unsupervised change detection in satellite images using principal component analysis and kmeans clustering. *IEEE Geoscience and Remote Sensing Letters* 6(4):772–776.

Congalton, R.G., and K. Green. 1999. *Assessing the Accuracy of Remotely Sensed Data: Principles and Practices* (Lewis Publishers, Boca Raton, FL), 137 p.

Congalton, R.G. 1991. A review of assessing the accuracy of classifications of remotely sensed data. *Remote Sensing of Environment* 37:35–46.

Dai, X.L., and S. Khorram. 1998. The effects of image misregistration on the accuracy of remotely sensed change detection. *IEEE Transactions on Geoscience and Remote Sensing* 36:1566–1577.

ERDAS. 2001. *ERDAS IMAGINE Tour Guides, ERDAS IMAGINE V 8.5* (ERDAS, Inc., Atlanta, GA).

ERDAS. 2010a. *ERDAS Field Guide* (Leica Geosystems GIS & Mapping, LLC, Norcross, GA).

ERDAS. 2010b. *IMAGINE Easytrace Tour Guide User's Guide* (ERDAS, Inc., Norcross, GA).

Goodchild, M.F., S. Guoqing, and Y. Shiren. 1992. Development and test of an error model for categorical data. *International Journal of Geographical Information Systems* 6:87–104.

Herold, M., X.H. Liu, and K.C. Clarke. 2003. Spatial metrics and image texture for mapping urban land use. *Photogrammetric Engineering and Remote Sensing* 69:991–1001.

Hester, D.B., H.I. Cakir, S.A.C. Nelson, and S. Khorram. 2008. Perpixel classification of high spatial resolution satellite imagery for urban land cover mapping. *Photogrammetric Engineering and Remote Sensing* 74:463–471.

Hester, D.B., S.A.C. Nelson, H.I. Cakir, S. Khorram, and H. Cheshire. 2010. High resolution land cover change detection based on fuzzy uncertainty analysis and change reasoning. *International Journal of Remote Sensing* 31:455–475.

Hord, R.M. 1982. *Digital Image Processing of Remotely Sensed Data* (Academic Press, New York), p. 256.

Intergraph Corporation. 2013. *ERDAS Field Guide* (Intergraph Corporation, Huntsville, AL), 792 p.

Jain, A.K. 1989. *Fundamentals of Digital Image Processing* (Prentice Hall, Englewood Cliffs, NJ), pp. 418–421.

Jensen, J.R. 2005. *Introductory Digital Image Processing*, 3rd ed. (Prentice Hall, Upper Saddle River, NJ), 316 p.

Khorram, S., S.A.C. Nelson, C.F. Van Der Wiele, and H.I. Cakir. 2016a. Processing and applications of remotely sensed data. In *Handbook of Satellite Applications*, 2nd ed. (Eds.) Pelton, J.N., Madry, S., and Camacho-Lara, S. (Springer-Verlag, New York).

Khorram, S., F.H. Koch, C.F. Van Der Wiele, S.A.C. Nelson, and M.D. Potts. 2016b. *Principles of Applied Remote Sensing* (Springer Science+Business Media, New York), p. 307.

Khorram, S., S.A.C. Nelson, H.I. Cakir, and C.F. Van Der Wiele. 2012a. Digital image acquisition: Preprocessing and data reduction. In *Handbook of Satellite Applications*, 2nd ed. (Eds.) Pelton, J.N., Madry, S., and Camacho-Lara, S. (Springer, New York), pp. 809–837.

Khorram, S., F.H. Koch, C.F. Van Der Wiele, and S.A.C. Nelson. 2012b. *Remote Sensing* in *Springer Briefs in Space Development* (Springer-Verlag, New York), p. 141.

Khorram, S., G.S. Biging, N.R. Chrisman, D.R. Colby, R.G. Congalton, J.E. Dobson, R.L. Ferguson, M.F. Goodchild, J.R. Jensen, and T.H. Mace. 1999. *Accuracy Assessment of Remote Sensing-Derived Change Detection* (American Society of Photogrammetry and Remote Sensing Monograph, Bethesda, MD), 64 p.

Lillesand, T., R. Kiefer, and J. Chipman. 2008. *Remote Sensing and Image Interpretation*, 6th ed. (John Wiley & Sons, New York), 763 p.

Mattison, D. 2008. Aerial photography. *Encyclopedia of Nineteenth-Century Photography*, Volume 1. (Ed.) Hannavy, J. (Routledge, New York), pp. 12–15.

Morgan, J.L., S.E. Gergel, and N.C. Coops. 2010. Aerial photography: A rapidly evolving tool for ecological management. *BioScience* 60:47–59.

Nelson, S.A.C., P.A. Soranno, and J. Qi. 2002. Land cover change in the upper Barataria Basin estuary, Louisiana, from 1972–1992: Increases in wetland area. *Environmental Management* 29(5):716–727.

Paine, D.P., and J.D Kiser. 2003. *Aerial Photography and Image Interpretation*, 2nd ed. (John Wiley & Sons, New York), 632 p.

Sabins, M.J. 1987. Convergence and consistency of fuzzy Cmeans/ISODATA algorithms. *IEEE Transactions on Pattern Analysis and Machine Intelligence* 9:661–668.

Shackelford, A.K., and C.H. Davis. 2003. A combined fuzzy pixel-based and object-based approach for classification of high-resolution multispectral data over urban area. *IEEE Transactions on Geoscience and Remote Sensing* 41:2354–2363.

Short, N.M. 2010. The Remote Sensing Tutorial [web site]. National Aeronautics and Space Administration (NASA), Goddard Space Flight Center. http://rst.gsfc.nasa.gov/.

Thomas, N., C. Hendrix, and R. Congalton. 2003. A comparison of urban mapping methods using high-resolution digital imagery. *Photogrammetric Engineering and Remote Sensing* 69:963–972.

Tou, J.T., and R.C. Gonzalez. 1977. *Pattern Recognition Principles* (Addison-Wesley, Readings, MA), p. 377.

Yu, Q., P. Gong, N. Clinton, G. Biging, M. Kelly, and D. Schirokauer. 2006. Object-based detailed vegetation classification with airborne high spatial resolution remote sensing imagery. *Photogrammetric Engineering and Remote Sensing* 72:799–811.

Index

Note: Page numbers followed by f and t refer to figures and tables respectively.

A

Accuracy assessment
  procedure, 277–284
  report generated from ERDAS
     IMAGINE, 284–286
  tool option, 278
Active imagery data, xv
Additional image analysis techniques,
  249
  application, 250–251
  impervious surface map creation,
    262–270
  land-only image creation, 251–257
  NDVI creation, 258–262
Advanced EasyTrace settings, 159f, 162f
Aerial photography/photogrammetry,
  xv, 2
Alarm Color, 182, 195
Anaglyphic glasses, 290f
Anaglyph images, 289
  for export creation, 298–300
Analysis Ready Data (ARD) images, 11
AOI layer option, 214f
Application
  additional image analysis
    techniques, 250–251
  assessing thematic classification
    accuracy, 274–275
  digital image processing application,
    52–53
  digital stereoscopy, 289–290
  image digitizing/interpretation,
    153–154
  package interface, 15f
  radiometric image enhancement,
    124–125
  spatial image enhancement, 148
Application of unsupervised
  classification
  approach, 177–182

  image subset of cloud-free areas
    creation, 172–176
  objectives, 169
  reference image subset creation,
    182–184
  required data, 169–172
Approximate True Color option, 179
ArcEditor, 39–40
ArcInfo, 40
ArcMap, 39
  connect to folder option, 41f
  data option, 42f
  Export Raster Data option, 44f
  image analysis option, 43f, 191f
ArcView, 39
Area of image subset, 174f
Area of Interest (AOI), 214, 237, 275
Assessing thematic classification
  accuracy, 273–274
  accuracy assessment procedure,
    277–284
  application, 274–275
  supervised classification recoding,
    275–277
Attribute table of grayscale
  unsupervised classification
  result, 182f

B

Bimodal histogram distribution, 132f
Block Image Path Editor, 301f, 302f
Breakpoint editor, 135f, 136f
Brightness contrast adjustment
  options, 133f

C

Camera calibration report, 100–101
Camera report fiducials order, 104f
Classes tab, 237

Index

Classification accuracy, 117
Classification interface, IMAGINE Objective, 236, 236f
Classification stage, supervised classification, 207–208
Classified/thematic image, 218, 218f
Color and definition scheme, supervised classification, 212
Color infrared (CIR) imagery, xv
Color ranges/band, xvii
Color scheme options, unsupervised classification, 179–182, 179f
Commission error, 273
Confusion tables, 273
Contents box, 56
Contrast adjust options/methods, 134f, 135f
Convolution filter, 147, 149f

D

Data, 1
"Data Type" (bit depth), 127
Delineating in stereo, 300–306
Diagonal elements, error matrix, 273
Dialog window, 16f
Digital raster imagery, xvii
Digital Stereo Model (DSM), 293
  setup as Relative Stereo Pair, 298f
Digital Stereo Pair making, 293–298
Digital stereoscopy
  Anaglyph Images for Export creation, 298–300
  application, 289–290
  Digital Stereo Pair making, 293–298
  Stereo Analyst module configuration, 290–292
Digitization, 153
Display properties of digitizing results, 165–166

E

EarthExplorer, 169
EarthExplorer web browser in coordinate section, 7f
  finding and downloading data, 6–27
  image scene footprint, 9f

Landsat 8 OLI/TIRS image, data search for, 170f
Landsat OLI/TORS data products, 8f
  product type information, 8f
  search criteria, 7f
  Sentinel-2 MSI data products, 9f
  show metadata/browse icon, 10f
Easytrace, 153
EasyTrace tool options, 158f
Edit | Create/Add Random Points option, 278
Electromagnetic radiation (EMR), xvii, 1
Electromagnetic spectrum (EMS), xv, 218, 249
ERDAS IMAGINE Graphic User Interface, xix–xx, 2–3, 56, 57f
  band combinations, 62–64, 63f, 63t
  data information, 62
  FCC display band combination, 64–65
  Help icon, 57f, 58
  multiple 2D viewer windows, 66
  New 2D View, 61f
  Online Help documentation, 58, 58f
  opening images, 60–62
  workspace preferences setup, 59, 59f
Error matrix, 273
Esri ArcMap ArcGIS
  for Desktop, 39–44
  unsupervised classification in, 190–199
Euclidean distance, 118, 118f, 121
Export creation, anaglyph images for, 298–300

F

False Color Composite (FCC), 173, 175, 213, 252, 277
  configuration, 159
  display band combination, 64–65
  image, xviiif
False Color image, 62
Feature extraction procedure, 237–244
Feature project setup procedure, 233–236
Fiducials numbering order, 104f
Field guides, 58

Index

325

Final output stage, supervised
classification, 209
Focal analysis filter, 147–148

G

Geographic Information System (GIS),
153, 275
Geometric distortions, 69
Geometric model, 72f, 82f
Georectification process, 69–70
image preprocessing, 70
Polynomial Regression, 71–85
Rubber Sheeting, 86–93
Georeferencing information, 11
Georeferencing process, 69
Geospatial Data Abstraction Library
(GDAL), 201
Graphic user interface (GUI), 181, 234
Grayscale color scheme options,
unsupervised classification, 180f
Gray scale image, 137f, 138f
Ground Control Points (GCPs), 87–89
input/reference, 89f
pointer locations, 105f
record, 74–80
selection, 109–113, 111f
tool dialog display, 112f
tool reference setup, 72, 73f
tool reference setup window, 98f
Ground truth, 273

H

Harmonic wave pattern, 1
Hot keys, digitizing, 160

I

Image classification, 167
Image data processing, 49–52
application, 52–53
in EarthExplorer, 53–56, 54f, 56f
ERDAS IMAGINE Graphic User
Interface, 56–67
packages, 50t–51t
Image Difference operation, 222
Image digitizing/interpretation
application, 153–154

display properties of digitizing
results, 165–166
overview, 153
polygon creation, 155–160
polyline creation, 161–165
Image enhancement process, 124
Image file naming conventions, 217t
Image layer transition swipe tool, 67f
Image metadata window, 128f
Image mis-registration, 117
Image preprocessing, orthorectification,
95–96
camera properties, 99–109, 99f
getting started, 96–99
selection GCPs, 109–113, 111f
Image subset options creation, 174f
Image-to-image change detection
comparisons
Image-to-image rectification, 70
IMAGINE Image (*.img) file format, 22f
IMAGINE Objective, 232
classification interface, 236, 236f
new project dialog window, 234f
save feature project, 235f
variable properties setup, 235f
workstation setup, 233f
Impervious surface map creation,
262–270
Import data option, 28f
Indices options, 259f
Infrared imagery, 2
Inquire cursor, 129, 129f
Iterative Self-Organizing Data Analysis
Technique (ISODATA)
approach, 167

J

JPEG 2000 (*.jp2) file format, 16f

K

Kappa statistic/Choen's Kappa, 285
K-means clustering approach, 168

L

Lake Wheeler location, 145f
Land cloud cover inspection, 170f

326 *Index*

Land cover
    category and color to apply, 176–177
    classification, unsupervised
        classification, 168t
Land-cover transition matrix, 227
Land-only image creation, 251–257
Landsat-7 or Landsat-8 from USGS
    dialog window, 33f
Landsat 8 FCC band combination
    selection, 65f
Landsat 8 OLI/TIRS, 4t, 8
    data import, 29f
    EarthExplorer data search for, 170f
    image, 61f
    IMAGINE Image (*.img) layer, 26f
    Layer Selection and Stacking dialog
        window, 24f
    layer stack operation, 25f, 172f
    Raleigh, North Carolina region, 13f,
        27f
    USGS EarthExpolorer dataset, 14t,
        31f, 34f
Landsat 8 subset image to Sentinel-2
    image comparison
    of extent, 183f
    swipe tool, 184f
Landsat satellite program, xix, 3
Language reference, 58
Layer Selection/Stacking dialog
    window, 18, 19f, 21f
Layer stack option, 17, 18f
"Layers to Colors" raster
    options, 23f
Line delineation of perennial stream,
    306f
Line delineation tool, 304f
Lookup tables (LUT), 129
    editor, 141f
    ERDAS IMAGINE and, 129–132
    to highlight water features, 144f
    versus raw image pixel file values,
        130t

M

Metadata icon, 62
Minimum-distance-to-means
    classification algorithm, 168
Model Maker toolbox option, 267f

Multi-band image, 18–19, 176f
Multipoint geometric correction dialog,
    75f, 98f, 103f
    control point color, 77f
    control point location, 76f
    total error, 78f
Multispectral data, xvii

N

NASA Satellite Data and Imagery
    Resources – Reverb, 3
National Agriculture Imagery Program
    (NAIP), 2
    classification setup, 259f
    creation, 258–262
New project dialog window, IMAGINE
    Objective, 234f
New shapefile layer option, 157f
NODATA warning, 100f
Non-directional edge filter, 147
Nonsystematic distortions, 69
Normalized Difference Vegetation
    Index (NDVI), 249

O

Object Based Image Analysis (OBIA),
    231–232
    application, 232–233
    feature extraction procedure,
        237–244
    feature project setup procedure,
        233–236
Object-oriented classification in ERDAS
    IMAGINE, 245–246
Omission error, 273
OpenGIS Simple Features Reference
    Implementation (OGR), 201
Open raster layer command, 15f
Operational Land Imager/Thermal
    Infrared Sensor (OLI/TIRS), 4,
    4t, 169, 209
Optical/infrared sensors, xv
Orthophotos, 95
    creation, 114–115
    viewing, 115, 116f
Orthorectification image preprocessing,
    95–113

Index

327

P

Parallelepiped classification algorithm, 208
Passive imagery data, xv
Pixel-based approach, 231
Pixels, xvi
Pixel values, 167
Polygon creation, 155–160
Polygon delineation tool, 304f
Polygon tool, 160f
Polyline creation, 161–165
Polynomial Regression, Schenk Forest using
  compute transformation matrix, 80
  display images, 71–74
  model properties, 74f, 80f, 82f
  record GCPs, 74–80
  resample image, 81–82, 81f
  verify rectification, 82–85, 84f, 85f
Positional accuracy assessment, 117–119
Pyramid file (*.rrd), 17
Pyramid layers option, 17f

Q

Quantum GIS (QGIS), 45
  application interface, 46f
  completed layer stack in, 47f
  raster data and multi-band image in, 45–47
  unsupervised classification in, 199

R

Radiometric image enhancement, 123–124
  application, 124–125
  concept, 124f
  ERDAS IMAGINE and LUT values, 129–132
  finer adjustments, 137–146
  performance, 125–132
  stretched/non-stretched images, 125–127, 127f
  stretch understanding, 127–129
Radiometric resolution, 123
Raleigh North Carolina region, 126f
Raster band combination, 125f

Raster clip in QGIS, 202f
Raster dataset, 1
Raster imagery, 49
Raster layer display options, 175f
Raster Object operators, 246
Raster to vector conversion, 242–243, 243f
Rectification display, 83f
Reference map coordinate system options, 88f
Reference map information, 73f, 99f
Remotely sensed data, 4–5
Remote sensing, xv
Report format definition, 79f
Resample dialog window, 114f
Reshape tool, 164f
Retriever box, 56
Road features digitization, 163f
Root mean square error (RMSE), 112f, 117–118, 118f, 120, 120f
Rubber Sheeting, Schenk Forest using
  compute transformation matrix, 89–90
  display images, 86, 86f
  ERDAS IMAGINE display, 93f
  GCPs, 87–89
  geometric correction tools, 87
  model properties dialog, 90f
  resample image, 90–91
  set geometric model, 87f
  swipe comparison of, 92f
  verify rectification, 91–93

S

Save feature project, IMAGINE Objective, 235f
Search library, 58
Sentinel-2 MSI bands *versus* layer stacked multi-band image, 19t
Sentinel-2 multispectral sensor, 5, 5t
Sentinel-2 TCC band combination selection, 65f
Sentinel Hub's Earth Observation (EO) Browser, 3
Set geometric model dialog, 97f
Signature file creation, supervised classification, 213–218
Single image options manipulation, 294f

328   *Index*

Spatial Analyst extension activation in ArcMap, 192f
Spatial image enhancement, 147–148
 3×3 filter initiation, 149
 5×5 high pass filter initiation, 150
 5×5 low pass filter initiation, 150
 application, 148
 edge detection filters, 149
 high pass filters, 147–148
 image degrade tool dialog window, 151
 low pass filters, 147–148
 non-direction edge filter tool dialog window, 150
Spectral/optical remote sensing, 1
Spectral Profile tool, 130, 131f, 252
Standard errors, 274
Stereo Analyst, 289
 module configuration, 290–292
 Stereo Mode setup, 292f
 toolbox option, 290f
 utility options, 291f
Straight-line distance, 118
Sum of off-diagonal elements, 273
Supervised classification, 207–209
 application, 209
 color and definition scheme, 212
 image difference operation, comparison using, 221–228
 initiating, 212–228
 recoding, 275–277
 required data, 210–211
 swipe tool, comparison using, 219–221
 tips, 218–219
Systematic distortions, 69

T

Thematic map, 167
Thematic Recode option, 275, 276f, 277f
Thermal Infrared Sensor (TIRS), 4, 4t
Three classification layers display, 266f
TIFF (*.tif) file format importing, ERDAS IMAGINE, 155f
Toggle viewer selectors, 106f
Toolbox option, Stereo Analyst, 290f
Total RMSE, 118, 120
Tour guides, 58

Training site polygons creation in active AOI layer, 215f
Training stage, supervised classification, 207
True Color Composite (TCC), 218, 277
True Color/natural color image, 62

U

Unimodal histogram distribution, 132f
United States Geological Survey (USGS), 169, 209, 250, 274
Unsigned 8-bit, 127–128
Unsupervised classification, 167–168
 application
 assigning class categories to, 184–188
 color scheme options, 179–182, 179f
 in Esri ArcMap ArcGIS, 190–199
 initiating, 176–177
 land cover classification, 168t
 options, 178f
 in QGIS, 199
Unsupervised KMeans image classification window, 202, 203f
USDA Natural Resources Conservation Service, 3
User guides, 58
User's accuracy, 274
USGS EarthExplorer, 3, 11, 12t, 32f
USGS Global Visualization Viewer (GloVis), 3
 cloud cover in, 37f
 data download option, 39f
 data selection, 37f
 finding/downloading data in, 34–39
 location of imagery, 36f
 metadata option review, 38f
 web browser Landsat Archive, 35f
USGS LandsatLook Viewer, 3
Utility options, Stereo Analyst, 291f

V

Variable properties setup, IMAGINE Objective, 235f
Vector Cleanup operators, 244
Vector layer loading, ERDAS IMAGINE, 156f

Index 329

Vector Object operators/processor, 243
Viewer swipe tool, 241f
Viewing properties, ERDAS IMAGINE, 166f
View | Select Viewer option, 278
Voronoi cells, 168

W

Warehouse Inventory Search Tool (WIST), 3
Water-only image mask, 249
Workstation setup, IMAGINE Objective, 233f